Srinivasa Reddy Yeduru

# Development of Microactuators Based on the Magnetic Shape Memory Effect

Schriften des Instituts für Mikrostrukturtechnik
am Karlsruher Institut für Technologie (KIT)
Band 21

Hrsg. Institut für Mikrostrukturtechnik

Eine Übersicht über alle bisher in dieser Schriftenreihe erschienenen Bände finden Sie am Ende des Buchs.

# Development of Microactuators Based on the Magnetic Shape Memory Effect

by
Srinivasa Reddy Yeduru

KIT Scientific Publishing

Dissertation, Karlsruher Institut für Technologie (KIT)
Fakultät für Maschinenbau

Tag der mündlichen Prüfung: 12. Juli 2013
Hauptreferent: PD Dr. rer. nat. Manfred Kohl
Korreferent: Prof. Etienne Patoor, Prof. Dr. rer. nat. Volker Saile

**Impressum**

KIT Scientific Publishing

Karlsruher Institut für Technologie (KIT)
KIT Scientific Publishing
Straße am Forum 2
D-76131 Karlsruhe

KIT Scientific Publishing is a registered trademark of Karlsruhe Institute of Technology. Reprint using the book cover is not allowed.

www.ksp.kit.edu



Print on Demand 2013

ISSN 1869-5183
ISBN 978-3-7315-0125-1

# Development of Microactuators Based on the Magnetic Shape Memory Effect

To obtain the academic degree of

**Doctor of Engineering**

From the faculty of Mechanical Engineering
Karlsruhe Institute of Technology

Approved

**Dissertation**

From

M.Sc. Srinivasa Reddy Yeduru

Date of examination: 12.07.2013

Main referee: PD Dr. rer. nat. Manfred Kohl

Co-referee: Prof. Etienne Patoor
Prof. Dr. rer. nat. Volker Saile

# Dedication

*The author would like to dedicate this research thesis to his parents (Nagi Reddy and Laxmidevi) and also to his teachers*

# Acknowledgements

This research work is performed at Institute of Microstructural Technology (IMT) in Karlsruhe Institute of Technology (KIT) in collaboration with IFW Dresden, Dresden and AdaptaMat Ltd., Finland. The present work receives funding from the Deutsche Forschungsgemeinschaft (DFG) and is part of the priority program SPP1239. One of the joys of completion of this research work is to look over the past journey and remember all the colleagues, friends and family who have helped and supported me along this long but fulfilling journey.

- First and foremost, I offer my sincere gratitude to my supervisor, PD Dr. Manfred Kohl who has supported me throughout my Ph.D study and research with his motivation, enthusiasm, and immense knowledge allowing me the room to work in my own way. My special thanks to Prof. Volker Saile, former director of IMT to put trust in me to continue my PhD study at the institute and to consider my request as one of the referees to my PhD thesis.
- Secondly I sincerely thank to Prof. Etienne Patoor, Director of Laboratory of Physics and Mechanics of Materials (LPMM), École Nationale Supérieure d'Arts et Métiers (ENSAM) group, France to accept my request to be as one of referees to my PhD thesis.
- I thank the collaborating research partners from IFW Dresden, Dr. Sebastian Fähler, Anja Backen and from AdpataMat Ltd, Dr. Alex Sozinov, Dr. Yossi Ezer who consistently supporting me by sending research samples for continuous flow of my research study.
- In my daily work I have been blessed with friendly and cheerful colleagues of 'Funcational Films' research group helped me regarding experimental and theoretical discussions to fulfill my research goals. I heartfully thank each of you for giving me such a nice experience to work like as a part of one family.

- I thank kindly Dr. Stephen Doyle who continuously accepting my request for doing experiments at ANKA (Angströmquelle Karlsruhe) synchrotron source and helping me to extract data and with further analysis
- I like to thank Dr. Christian Kübel, Dr. Di Wang and Dr. Torsten Scherer for providing access to perform experiments in High Resolution Transmission Electron Microscopy (HR-TEM) at Karlsruhe Nano-Micro Facility (KNMF) and also for their fruitfull discussions while analyzing TEM data at Institute of Nanotechnology (INT), Karlsruhe Institute of Technology.
- Specially I thank Dr. Klaus Feit from IMT for helping me in performing Tensile experiments and also Differential Scanning Calorimetry (DSC) measurements
- I also thank Mr. Ulrich Klein from IMT helped me to my software requests through his friendly involvements.
- Thanks too all employees of IMT who directly or indirectly support to my research work and my special thanks to Prof. Jürg Leuthold, Prof. Andreas Guber, Dr. Jürgen Mohr, Dr. Matthias Worgull for their important suggestions and Dr. Dieter Maas, Dr. Holger Moritz for their help regarding administrative matters.

Finally, I thank sincerely my parents, wife, sister, brother and my friends for inspiring me throughout my journey while staying in Germany.

November 7, 2013 Srinivasa Reddy

# Abstract

The giant magneto-strain effect in Ni-Mn-Ga alloys is particularly attractive for actuator applications in micro-and nanometer dimensions as it enables contact-less control of large deflections in the order of 10%. Two different approaches are being pursued to develop ferromagnetic shape memory microactuators based on magnetically induced reorientation of martensite variants. One is by following a top-down approach of thickness reduction of bulk Ni-Mn-Ga single crystal foil specimens from 200 to 30μm. Second one is the development of freestanding epitaxial Ni-Mn-Ga thin film actuators in a bottom-up manner by magnetron sputtering, substrate release and integration technologies. To observe large deflections of Ni-Mn-Ga microactuators in microsystems, the actuator material should be exhibiting low mechanical twinning stress and large magnetic anisotropy. In addition, design rules and boundary conditions for operating the Ni-Mn-Ga actuator material are having significant importance for evolution of performance characteristics.

Ni-Mn-Ga foils prepared by electric discharge method reveal thickness-dependent microstructural, phase transformation and mechanical properties. The continuous increase in twinning stress while decreasing the thickness of the foil from 200μm to 30μm (least thickness) is observed in stress-strain characteristics. This behavior is studied by the influence of defect surface layer on the twin boundary mobility. Microstructural investigations performed on these defect surface layers reveal high fraction of lattice dislocation and stacking faults which are induced by the foil preparation method. After thermo-mechanical training, 200μm thick Ni-Mn-Ga foil shows a partial MIR up to 1% above a critical magnetic field of 0.2T. The effect of thermo-mechanical training in thinner foils below 200μm thickness seems to be less significant due to remaining large fraction of defect density.

Two single crystalline Ni-Mn-Ga foils of 200μm and 30μm are considered to study the linear actuation characteristics. These foils are prepared by abrasive wire sawing method by minimising the severe surface damage. Ni-Mn-Ga foils with 200μm thickness are exhibiting 10M modulated crystal structure at RT. The low twinning stress of 1.2MPa for 200μm thick Ni-Mn-Ga foil is observed in stress-strain measurements. Large reversible actuation strain of 5.9% above a critical magnetic field of 0.12T for 200μm and 30μm foils is reported. Tensile load-dependent magneto strain characteristics are investigated for 200μm and 30μm thick Ni-Mn-Ga foils. The optimum tensile stress to enable full linear actuation cycles in Ni-Mn-Ga foils is investigated. Ultimately, full linear actuation cycles with a maximum actuation stroke of 2.2% upon biasing with constant tensile load of 1.5MPa in a magnetic field are generated based on Ni-Mn-Ga foil actuators of 200μm thickness.

Design rules are developed for optimum fixation of the Ni-Mn-Ga foil actuator; it is suggested to fix the Ni-Mn-Ga foil at the face side of the cross-section in order to avoid blocking of twin boundaries due to fixation. In addition, homogeneous magnetic field distribution subjected to Ni-Mn-Ga foil is required in order to gain high reversible actuation strains. Material inhomogenities in Ni-Mn-Ga foil have to be taken into account since they can limit the reversible actuation strain. To increase the actuation stroke in Ni-Mn-Ga foil actuator coupling with external bias spring mechanism is suggested by considering the optimum prestress, prestrain and spring constant of the bias tensile spring.

A novel Cr-sacrificial layer technology is developed to release epitaxial Ni-Mn-Ga films from the MgO substrate. Ni-Mn-Ga films are exhibiting co-existing 14M and NM martensites at room temperature. Stress induced variant reorientation in 1μm thick freestanding Ni-Mn-Ga film reveals a large super plastic strain of 12% and twinning stress of 25MPa along the [100] direction of the Ni-Mn-Ga unit cell. After complete variant reorientation, the resulting crystal structure is identified to be detwinned NM martensite. In addition, tensile experiments performed on [110] direction of Ni-Mn-Ga unit cell reveals a plastic strain of 4% with high twinning stress of 30MPa. The orientation dependent twinning stress and strain in freestanding Ni-Mn-Ga films are distinguished by the different orientations of external tensile load with respect to twinning planes.

Two stage phase transformation behavior in 2μm thick freestanding Ni-Mn-Ga films is identified due to complex martensitic transformation between 14M and NM martensites. Temperature dependent magnetization, electrical resistance, x-ray diffraction and stress-strain measurements suggest that a first order transformation occurs in temperature regime (I) of 40°-80°C and followed by an intermartensitic transformation in temperature regime (II) of 140°-160°C. Finally, thermo-mechanical training methods are investigated on freestanding Ni-Mn-Ga films. Training is performed on freestanding Ni-Mn-Ga films by applying a constant stress of 2 and 6MPa along the tensile loading direction. Five temperature cycles are performed between martensite finish and austenite start temperatures. After thermo-mechanical training, a drastic decrease of twinning stress below 3% strain is observed. This is attributed to a larger fraction of 14M martensites that are partially aligned with their long axis along the tensile loading direction.

# Kurzfassung

Der magneto-Dehnungseffekt in NiMnGa Legierungen ist für Aktoranwendungen im Mikro- und Nanometerbereich besonders attraktiv, erlaubt er doch die berührungslose Kontrolle großer Auslenkungen im Bereich 10%. Zwei verschiedene Ansätze zur Entwicklung ferromagnetischer Formgedächtnis-Mikroaktoren auf Basis der magnetisch-induzierten Reorientierung martensitischer Varianten werden verfolgt. Ein Ansatz nutzt top-down Prozesse um die Foliendicke von NiMnGa Einkristallen von 200 auf 30μm zu reduzieren. Der zweite, bottom-up basierte Ansatz, besteht aus der Entwicklung freistehender, epitaktischer NiMnGa-Dünnfilmaktoren durch Magnetronsputtern, Substratabhebe- und Integrationstechnologien. Um große Auslenkungen für NiMnGa Mikroaktoren in Mikrosystemen zu erhalten, sollte das Aktormaterial kleine mechanische Zwillingsspannungen und eine große magnetische Anisotropie aufweisen. Zusätzlich haben Designregeln und Grenzbedingungen für den Betrieb der NiMnGa-Aktormaterialien eine besondere Bedeutung für die Weiterentwicklung der Leistungscharakteristiken.

Mikrostruktur, Phasentransformation und mechanische Eigenschaften von NiMnGa-Folien, die mit der Methode der Funkenerosion hergestellt wurden, zeigen ein dickenabhängiges Verhalten. Der stetige Anstieg der Zwillingsspannung bei abnehmender Foliendicke von 200 zu 30μm (geringste Dicke) wird in Spannungs-Dehnungs-Charakteristiken beobachtet. Dieses Verhalten wird dabei über den Einfluss von beschädigten Oberflächenschichten auf die Zwillingsgrenzenbeweglichkeit untersucht. Mikrostrukturelle Untersuchungen an diesen Oberflächenschichten ergeben einen hohen Anteil an Versetzungen und Stapelfehlern, die auf die Herstellungsmethoden zurückzuführen sind. Nach einer thermo-mechanischen Vorbehandlung zeigen die 200μm dicken NiMNGa-Folien eine teilweise MIR bis zu 1% über einem kritischen Magnetfeld von 0,2T. Der Einfluss der thermo-mechanischen Vorbehandlung bei dünneren Folien unter 200µm scheint aufgrund der größeren Defektdichte weniger bedeutend zu sein. An zwei einkristallinen NiMnGa Folien der Dicke 200 und 30µm wird die lineare Aktuierungscharakteristik untersucht. Diese Folien werden mit der Methode des abrasive wire sawing hergestellt um schwere Oberflächenschäden zu minimieren. NiMnGa Folien mit 200µm Dicke weisen bei Raumtemperatur eine 10M modulierte Kristallstruktur auf. Die geringe Zwillingsspannung von 1.2MPa für die 200µm dicke Folie wird in Spannungs-Dehnungs Messungen beobachtet. 200 und 30µm dicke Folien zeigen eine große, reversible Aktuierungsdehnung von 5,9% über einem kritischen Magnetfeld von 0.12T. Zugbelastungsabhängige magneto-Dehnungs Charakteristiken werden für NiMnGa Folien der Dicke 200 und 30µm untersucht. Die optimale Zugbeanspruchung, um volle, lineare Aktuierungszyklen zu erhalten, wird untersucht. Zuletzt werden volle, lineare Aktuierungszyklen mit einem maximalen Aktuierungshub von 2,2% bei einer Vorspannung mit einer konstanten Zugspannung von 1.5MPa in einem Magnetfeld realisiert. Diese basieren auf NiMnGa Folienaktoren mit 200µm Dicke.

Es werden Designregeln für eine optimale Fixierung der NiMnGa Folienaktoren entwickelt; es wird vorgeschlagen die NiMnGa Folien an der Stirnseite des Querschnitts zu fixieren um eine Blockierung der Zwillingsgrenzen aufgrund der Fixierung zu vermeiden. Zusätzlich ist es notwendig, die Folien einem homogenen magnetischen Feld auszusetzen um hohe, reversible Aktuierungsdehnungen zu erhalten. Materialbedingte Inhomogenitäten in NiMnGa Folien müssen berücksichtigt werden da sie die reversible Aktuierungsdehnung begrenzen können.Um den Aktuierungshub zu erhöhen wird die Verwendung einer externen vorgespannten Feder mit optimierter Vorspannung und Federkonstante empfohlen. Es wurde eine neuartige Cr-Opferschicht-Technologie entwickelt, mit deren Hilfe sich epitaktische NiMnGa Filme von einem MgO-Substrat lösen lassen. In NiMnGa Filmen existieren bei Raumtemperatur gleichzeitig 14M und NM Martensite. Die spannungs-induzierte Reorientierung der

Varianten in 1µm dicken freistehenden NiMnGa Filmen zeigt eine große superplastische Dehnung von 12% und Zwillingsspannungen von 25MPa entlang der [100] Richtung der NiMnGa Einheitszelle. Nach einer kompletten Reorientierung der Varianten lässt sich der resultierende Kristall als entzwillingter NM Martensit identifizieren. Zugversuche, die entlang der [110] Richtung der Einheitszelle durchgeführt wurden, zeigen eine plastische Dehnung von 4% mit einer hohen Zwillingsspannung von 30MPa. Mit Zugbelastungen, die unterschiedlich zu den Zwillingsebenen orientiert sind, werden die richtungsabhängigen Zwillingsspannungen und –dehnungen in freistehenden NiMnGa Filmen auseinandergehalten.

Ein zwei-Stufen Phasentransformationsverhalten in 2µm dicken, freistehenden NiMnGa Filmen kann aufgrund der komplexen martensitischen Transformation zwischen 14M und NM Martensit identifiziert werden. Messungen der temperaturabhängigen Magnetisierung, des elektrischen Widerstandes, der Röntgenbeugung und des Spannung-Dehnung Verhaltens legen eine Phasentransformation ersten Grades im Temperaturbereich (I) 40°-80°C und eine intermartensitische Transformation im Bereich (II) 140°-160°C nahe. Zuletzt werden thermo-mechanische Vorbehandlungmethoden an freistehenden NiMnGa Filmen untersucht. Die Vorbehandlung wird an freistehenden NiMnGa Filmen durchgeführt, indem eine konstante Spannung von 2 und 6MPa entlang der Zugbelastung angelegt wird. Zwischen der Martensit-End und der Austenit-Start Temperatur werden fünf Temperaturzyklen durchgeführt. Nach der thermo-mechanischen Vorbehandlung wird ein deutlicher Abfall der Zwillingsspannung unter 3% Dehnung beobachtet. Dies wird auf einen größeren Anteil an 14M Martensit zurückgeführt, der mit seiner langen Achse teilweise entlang der Zugbelastung ausgerichet ist.

# Contents

# Chapter 1

# Introduction

## 1. 1 Motivation

Actuators are the functional elements, which connect the information processing part of an electronic control system with a technical or nontechnical (e.g. biological) process. Actuators affect processes in its environment, through flow of information, energy or matter. Actuators of macroscopic size have wide applicability in industrial, automotive, aerospace, robotics, space and medical fields. For most of the applications the energy input of an actuator is controlled electrically and the output is generally in the form of mechanical working potential. The essential part of "electro-mechanics" is the generation of mechanical energy from electrical energy. Usually there are different ways to perform this; conventionally, most of the devices used for such energy conversion are the electromagnetic machines that use magnetic fields. On the other hand smart materials can be used as actuators and sensors because of their multifunctional behavior. In addition, smart materials exhibit energy conversion, for instance, between mechanical energy and electrical energy. Some of the widely recognized smart materials include piezoelectric, magnetostrictive, and shape memory alloys [1].

The demand on microactuators tends to increase as a direct consequence of the progress in the field of microelectronics and its development is associated with high requirements for materials, design methods and technologies. Microactuators are mostly based on planar mechanical structures with very small dimensions which are produced with the help of lithographic procedures

and anisotropic etching techniques [2]. Scaling macroactuators down to microactuators may affect the performance of the total system drastically. For instance, one has to keep in mind that thin films, which are used for the batch fabrication of microactuators, often have different physical properties compared to bulk materials including intrinsic stress induced by high temperature processing and deposition [3]. Compared to macro and mini-actuators, microactuators allow completely new mechanical designs with enormous design freedom. Thus, microactuators may be combined with microelectronics, photonics and biofluidics onto a single substrate to perform various functions.
Microactuators have an extensive applicability in handling of small amount of fluids in medical, biological, and chemical engineering, as well as the execution of switching or positioning functions in information technology, safety technology, and robotics [4]. For microactuator with large displacement, many different principles for force generation are used. The right choice of the actuation principle depends on many factors, such as structural dimensions, technology, response time, force and the torque as a function of displacement and the maximum power consumption.

Force can be generated by the following different principles:

1. External forces which are generated in the space between stationary and moving parts using for example thermopneumatic effects, and electrostatic or magnetic fields.
2. Intrinsic forces in solid state actuators are generated by special materials having actuation capabilities including piezoelectric, thermo-mechanical, electrostrictive, magnetostrictive and shape memory effects.

However, the advantages and disadvantages of a given type of microactuator depend on various factors such as scaling, design, fabrication, material and physical properties. Among several actuation principles, the actuation based on shape memory alloys gained a large interest due to (unique) favorable down scaling behavior along with high energy densities as compared with conventional actuators. In addition, shape memory alloys belong to a special class of "smart materials" because of their multifunctional behavior, besides showing actuation functions they are also used in temperature sensing, electrical, or structural functions. This enables simple and compact designs with multifunctional features. Due to these reasons, shape memory alloys are particularly interesting for applications in micro and nano-scale dimensions. However, the shape memory effect involves temperature-induced phase transformation with low thermodynamic efficiency. Therefore, bulk (macro) actuators have a characteristic disadvantage of low heat transfer rates. However, for shape memory microactuators this disadvantage is less severe as they gain high frequencies due to reduced thermal mass.

Recently, giant magnetostrictive materials that change their shape in magnetic fields attracted attention from scientific community. Magnetostriction can be described as a deformation of a body in response to a change in magnetization of the material [5]. There are many different materials (Ni, Co, Fe, $Fe_3O_4$, $Tb_{0.5}Zn_{0.5}$, Terfenol-D, FeSiBC) that can change their shape based on this effect [6]. Due to giant magnetostriction effect, these materials can exhibit large strains (0.24% for Terfenol-D [6]) and are therefore used as smart materials. A new class of magnetostrictive materials with giant magnetostriction effect has been discovered by Ullakko et al. [7] called Magnetic Shape Memory Alloys (MSM) or Ferromagnetic Shape Memory Alloy (FSMA) materials. The modification of microstructure and shape of solid material by an external magnetic field is described as magnetic shape memory (MSM) effect [8] or twin boundary induced magnetostriction [9].

The main difference between the classical magnetostriction and MSM effect is the magnitude of strain. Magnetostriction occurs in a material by change in the orientation of magnetic domains towards external magnetic field, where as MSM effect occurs in a material by change in the orientation of crystallographic axis towards magnetic field. Thus, magnetostriction strain is much smaller (~25 times) than MSM strain. The direct and inverse magnetic shape memory effects cause magnetic field induced super elasticity, which is the large magnetically induced recovery of mechanically induced deformation [10]. The thermal shape memory effect can also be observed in MSM materials when the material is heated above the martensite-austenite transformation temperature. However, MSM effect itself occurs in the martensite phase and does not need a phase transformation. Hence the magnetic field triggered MSM effect is considerably faster than the thermal shape memory effect.

Much of the research work on MSM alloys is based on Ni-Mn-Ga Heusler alloys because of their outstanding material properties such as martensite crystal structure, transformation temperatures and magnetocrystalline anisotropy. Hence, in Ni-Mn-Ga alloys large strains up to 10% by MSM effect are reported. However, the amount of generated strain is not the only important parameter for a smart material. The generated output stress and remote controllability along with high frequency and the needed external components for material operation are also essential considerations when determining the benefits of different type materials.

MSM materials exhibit different coupling effects being of interest for actuator, sensor and energy harvesting [11] applications at micro and nanometer dimensions [12]. The antagonism of ferromagnetic and shape memory effect makes use of electro-thermo-mechanical coupling [13], where magnetic field can induce mechanical deformation using magneto-mechanical coupling [14] and also phase transformation can be triggered by magnetic field through magneto-thermal coupling [15]. Additionally, using the Joule effect (electro-thermal actuation) and Ampere's law (electro-dynamic generation of a magnetic field) one could introduce more coupling effects in MSM microactuators. This is illustrated by a schematic as shown in Fig. 1.1.

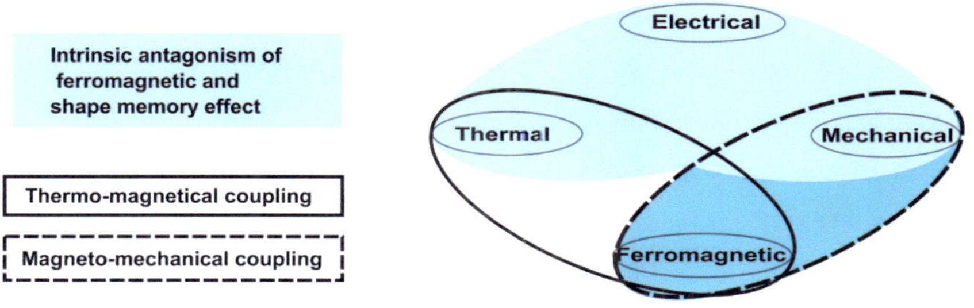


**Figure 1.1:** Schematic of different coupling effects in MSM materials of interest for actuator applications.

In combination, different coupling effects allow to enable contactless control of large deformations for MSM microactuators. Therefore, development of microactuators based on MSM material is highly motivated for microsystem applications.

## 1.2 Objective

For development of a MSM microactuator, the following aspects require comprehensive investigation regarding materials, engineering and technology. From the materials point of view, epitaxial Ni-Mn-Ga films or single crystalline Ni-Mn-Ga foils are considered as right candidates to obtain large strains, because significant strains in bulk MSM materials to date is only obtained in single crystals. However the fabrication of Ni-Mn-Ga foils or films showing similar martensitic microstructure as bulk Ni-Mn-Ga alloys turn up lot of technical challenges. In addition, by decreasing one dimension the surface to volume ratio increases, which may have high influence on the mechanical properties, which is crucial for MSM actuation.

For Ni-Mn-Ga foils, in the literature there is only one paper that shows qualitative results on magneto strains that were calculated theoretically through magnetization measurements [16]. Hence, for Ni-Mn-Ga foils, there are several unknown issues exist mainly concerning the mechanical properties, actuation characteristics in a magnetic field, boundary conditions and design rules. In addition, MSM device performance based on Ni-Mn-Ga foil actuator still has to be investigated.

Since Ni-Mn-Ga foils are having certain technological limitation about their minimum thickness, hence alternatively through bottom-up approach, epitaxial Ni-Mn-Ga films are investigated to develop microactuators. Out of several preparation methods, sputtering is identified to deposit reproducible epitaxial films on various substrates. Recent efforts mainly concentrate on fabricating epitaxial Ni-Mn-Ga films on a single crystalline MgO substrate. However after deposition, it is found that Ni-Mn-Ga films on a rigid substrate do not show large magneto strains. Hence, the Ni-Mn-Ga films have to be released from the substrate through a reliable technological approach. After releasing from the substrate, the materials and mechanical properties for freestanding epitaxial Ni-Mn-Ga films needs to be investigated. Furthermore, different training methods to improve mechanical performance of Ni-Mn-Ga films have to be investigated as well. Despite of this, due to easily buckling nature of freestanding Ni-Mn-Ga films, additional boundary conditions and technological approaches needs to be studied to enable actuation in freestanding Ni-Mn-Ga films.

In this research thesis, foils of Ni-Mn-Ga are prepared by top-down approach. Different technical ways to prepare Ni-Mn-Ga foils are discussed and their influence on the Ni-Mn-Ga foil material, mechanical and actuation characteristics in a magnetic field are investigated. Training is performed on Ni-Mn-Ga foils to improve the mechanical properties in order to observe MSM actuation in magnetic field. In addition, different boundary conditions are identified and their influence on the actuation strain in a magnetic field are discussed. The reversible linear actuation in Ni-Mn-Ga foils is studied and a miniature actuator is demonstrated. Different design rules for Ni-Mn-Ga foil actuator are discussed and established. Finally, the repeated cycles of linear actuation based on reset bias mechanism are demonstrated.

Epitaxial Ni-Mn-Ga films are deposited on MgO substrate with Cr as a sacrificial layer. Hence different fabrication technologies are studied to release the films from the substrate. Freestanding Ni-Mn-Ga films are investigated towards understanding their mechanical properties, magnetic properties coupled with material properties. In addition, mechanical properties at different transformation temperatures are investigated in order to understand the complex phase transformation in freestanding epitaxial Ni-Mn-Ga films. Finally, the effect of thermo-mechanical training on the mechanical properties is investigated.

## 1.3 Outline of the thesis

This thesis begins with a brief overview on the field of actuators considering the high impact of ferromagnetic shape memory materials as microactuators in microsystems. Chapter-2 begins with an overview of shape memory and ferromagnetic shape memory alloys in terms of shape memory phenomenon, mechanism involved and theoretical background on different kind of martensite crystal structures. The material development and preparation routes of Ni-Mn-Ga foils and Ni-Mn-Ga thin films are also presented. The material, magnetic and mechanical characterization techniques are briefly explained in chapter-3. The various fabrication methods are discussed in chapter-4 to achieve the freestanding Ni-Mn-Ga films without the role of rigid substrate constraint. Furthermore, various design schematics of the Ni-Mn-Ga film based microactuators are explained and discussed towards application point of view.

The results focusing on characterization of Ni-Mn-Ga foils prepared by using top-down approach are presented in Chapter-5. Their performance characteristics and thereby dependence on the counteracting tensile load are discussed in details for different boundary conditions. The thickness-dependent performance of linear actuators is also discussed. The novel process flow to release the Ni-Mn-Ga films with Cr sacrificial layer are discussed and then freestanding Ni-Mn-Ga films are investigated towards understanding the relation between the microstructure and mechanical properties in chapter-6. In addition, temperature-dependent microstructural investigations and thermo-mechanical training results are presented towards understanding the freestanding Ni-Mn-Ga films to enable low twinning stress and ultimately magnetic field induced actuation.

Finally, conclusions from this research followed by suggestions to future work are given in chapter-7. The outlook of demonstrating FSMA microactuator is presented by suggesting the wide applicability of ferromagnetic shape memory microactuators in optics, automotive, energy harvesting etc.,

# Chapter 2

# Background

Shape memory alloys (SMAs) show fascinating thermo-mechanical properties such as the shape memory effect and superelasticity. They are able to recover from large permanent deformations by slightly increasing their temperature or from large strains upon loading and unloading the material. The physical mechanism responsible for such a peculiar behavior is a martensitic transition from an open symmetry phase to a close-packed structure. From a technical point of view, these materials are very attractive since they can function as sensors as well as actuators. Considerable loss of potential efficiency of thermally controlled shape-memory actuators is due to the slow mechanical response to temperature changes. Hence magnetic shape memory (MSM) alloys are promising candidates to overcome such a problem by opening up the possibility of magnetic control of shape memory effect.

## 2.1 Shape memory alloys

The essential condition for a material to belong to the group of shape memory alloys is that the so-called thermo-elastic martensitic transformation of the crystal lattice takes place. In this phase transformation, the crystal structure changes from the parent martensite phase having tetragonal or orthorhombic crystal structure to a high symmetry cubic austenite phase. The schematic shown in Fig. 2.1 explains the conventional thermal shape memory effect. If a shape memory sample in the martensitic state gets deformed, its original shape through phase transformation can be induced by increasing the temperature of the material above a certain critical point. This phase transformation is diffusionless, i.e., the chemical composition of the material does not change and only structural rearrangements occur in the crystal lattice. Thus, this process is highly reversible. The reverse phase transformation starts by the annihilation of twin variants in the parent martensite phase.

The origin of different twin variants separated by twin boundary is illustrated in Fig. 2.2. Since the lattice constants in a martensite material having tetragonal crystal structure are different and also orientated in different crystallographic direction, these differently orientated martensite regions are distorted with respect to the surrounding lattice. This leads to the local strain either by elongation or contraction along the preferred crystallographic direction.

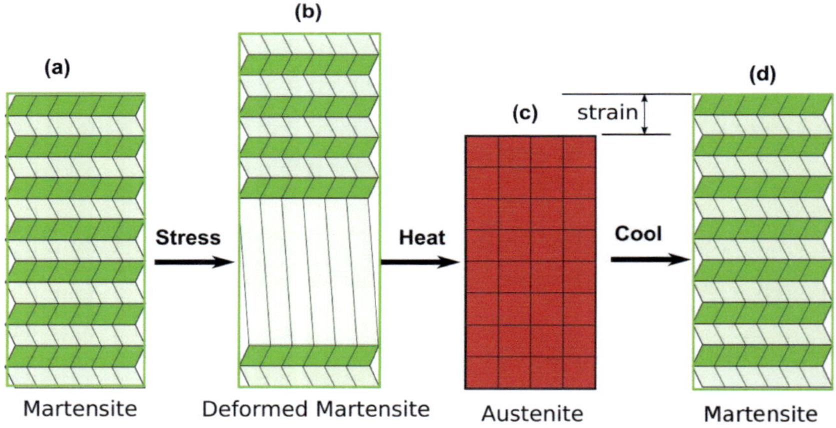


**Figure 2.1:** Schematic illustration of the temperature-driven two way shape-memory effect. (a) The parent martensite phase, (b) deformation of the sample by an external stress, (c) heating the sample to high temperature to induce austenite phase, (d) the original shape is restored upon cooling the sample from the high temperature austenite.

The martensitic planes may slip plastically and an ensemble of twin variants or domains separated by mobile twin boundaries is formed. The mobile twin boundaries can also be regarded as a mirror-like stacking fault of neighboring atomic layers. Usually, the motion of the twin boundaries is the easiest way for the sample to deform since in contrary to slipping this process involves breaking fewer chemical bonds in the lattice and provides the clearest explanation to the macroscopic shape change [17].

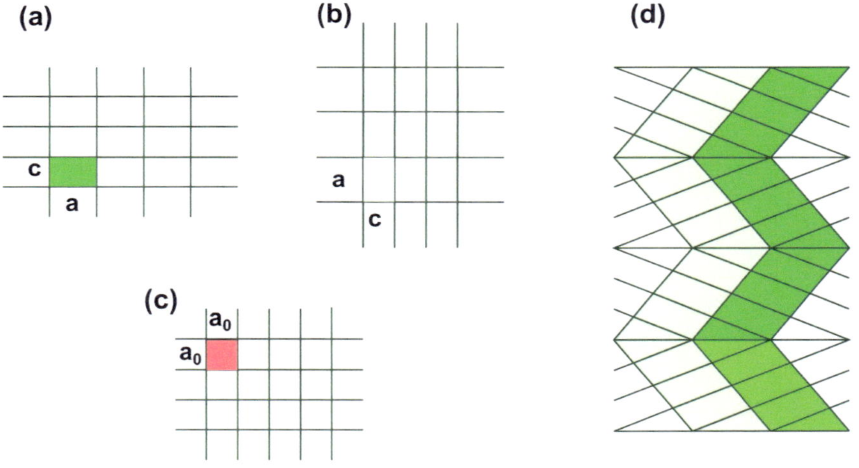


**Figure 2.2:** Formation of twin structure in two dimensions for a NiTi material. (a), (b) Two different martensitic variants for a tetragonal crystal structure, (c) final austenite phase for cubic structure, (d) regions of martensitic variants separated by twin boundaries.

In the temperature driven shape memory effect, the transformation begins at a temperature $T_{As}$ and ends at $T_{Af}$. The start and finish temperatures for the reverse process are $T_{Ms}$ and $T_{Mf}$, respectively. This transformation is hysteretic implying that the reverse process occurs at slightly different temperatures than the original transformation as shown in Fig. 2.3. The width of this thermal-hysteresis loop depends on the amount of elastic and surface energies stored during the transformation and, generally the narrower the loop the easier it is to switch from one phase to another phase upon heating or cooling. Typically, the thermal hysteresis is of the order of 10-20ºC, but for some alloys it can be as much as 50ºC or even more [18]. The schematic shown in Fig. 2.3 explains the one-way shape memory effect, restoring from previous shape changes induced by manipulation of martensite microstructure. It can even switch between two defined shapes by undergoing martensitic and reverse transformations (two-way SME). Other unusual mechanical properties of SMAs include superplasticity, rubberlike behavior etc. [18]. One-way shape memory effect was discovered in 1951 in a gold alloy containing 47.5 at.% of Cadmium but became a field of active interest only after similar behavior had been observed in Ni-Ti alloys [18, 19].

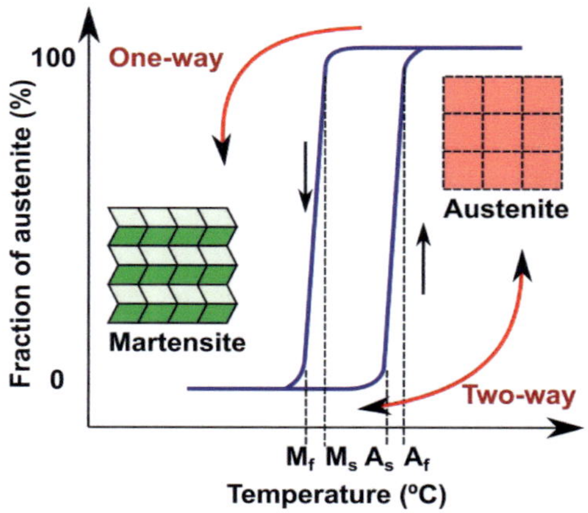


**Figure 2.3:** Temperature hysteresis using shape memory effect, where low temperature phase is twinned martensite and high temperature phase is symmetric austenite.

The twin boundaries in the martensitic phase are highly mobile, which makes it possible to deform the sample easily. During the deformation process, the volume fraction of the twin variant exhibiting the largest strain in the direction of the external force grows at the expense of other variants. As the stress applied to the sample reaches a certain critical value, the twin boundaries start to move and as a result, larger strains are achieved at an approximately constant stress. The twin-boundary motion preserves the topological structure of the lattice, which also explains why in the temperature-driven shape memory effect it is possible to return to the original state upon cooling. The observed strains can be large, up to 8% in NiTi alloys, but the operating frequency is limited to 100 Hz due to mechanical or thermal nature of the phenomenon [3, 20]. Therefore, several attempts have been made to discover materials with a fast response to the external force and at a high operating frequency.

## 2.2 Ferromagnetic shape memory alloys

Ferromagnetic shape memory alloys (FSMA) or magnetic shape memory (MSM) alloys are a new class of materials, which are receiving considerable interest from scientific community and industry. The coupling between the magnetic and structural degree of freedom confer unique magneto-mechanical properties to this material. What makes the ferromagnetic shape memory alloy having distinct martensite microstructure from ordinary shape memory alloys is that the martensite microstructure can be tuned by an applied magnetic field as shown in Fig. 2.4. This can lead to large strains in the material. Additionally, these changes can be 'remembered' by the martensite microstructure.

Ullakko and his co-workers first discovered the ferromagnetic shape memory effect on Ni-Mn-Ga alloys in 1996 [7, 21]. They demonstrated a 0.2% strain in stoichiometric $Ni_{50}Mn_{25}Ga_{25}$ specimen induced by 0.8T magnetic field and they interpreted the effect as a result of rearrangement of martensite microstructure.

Since the initial demonstration of the possible magnetic field-induced actuation, the maximum output strain achieved has been increased from the 0.2 % to 6%, which is close to the theoretical maximum by the tetragonal crystal structure [22, 23]. Later, the large magnetic field induced strain of 10 % in Ni-Mn-Ga MSM alloy has been observed by Sozinov et al. in 2002 [24] and Müllner et al in 2004 [25]. The MSM effect is also observed in Ni-Mn-Ga-Fe [26], Fe-Pd [27], La-Sr-$CuO_4$ [28], Fe-Ni-Co-Ti [29, 30], Fe-Pt [31], Co-Ni-Ga [32], Ni-Mn-Al [33] and Co-Ni-Al alloys [34]. Despite of being many magnetic shape memory alloys, $Ni_2MnGa$ heusler alloys are being extensively investigated because of their unique magneto-mechanical properties [35].

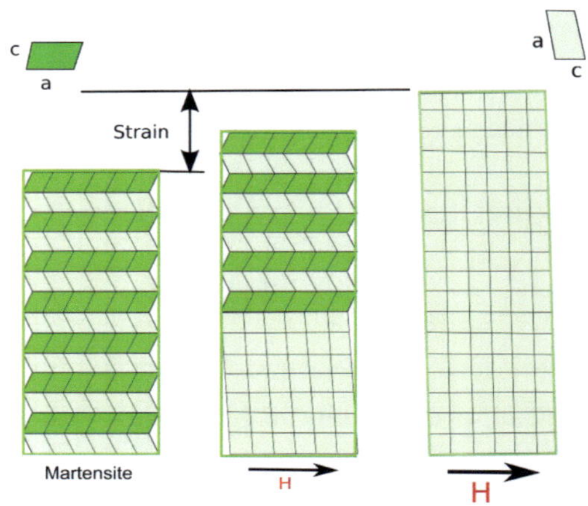


**Figure 2.4:** Left: A FSMA material consisting of a martensitic variant structure at zero field. Center: When a magnetic field orthogonal to the orientation of the variant is applied, a new martensitic variant (light green colour) oriented parallel to the field. Right: as the field strength is increased, the proportion of this new variant increases and, simultaneously the shape is elongated.

In the magnetic shape memory alloys ($Ni_2MnGa$), the crystallographic structure of the low temperature martensite exhibits tetragonal symmetry. Fig 2.5 depicts three possible tetragonal variants (variant 1, variant 2 and variant 3) along different crystallographic orientations with saturation magnetization direction.
Each of the martensitic variants are magnetized along a preferred crystallographic direction, named the magnetic easy axis, which in this case is

aligned with the short edge ***c*** of the tetragonal unit cell, where it must be understood that the magnetization of a single unit cell is an idealized concept. The magnetization (**M**) can be oriented in either the positive of negative easy axis direction. With this configuration of orthogonal preferred directions of magnetization, it is evident that the application of an external magnetic field can be used to favor selected variants over others, since an alignment of the magnetization and the magnetic field inside a magnetic material is energetically desirable. In the context of MSM alloys, this process is referred as magnetic field induced reorientation (MIR) of martensitic variants, which is the microstructural mechanism causing the macroscopic magnetic shape memory effect.

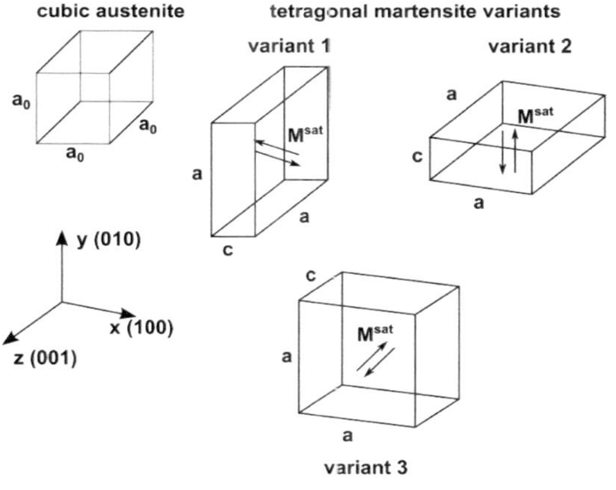


**Figure 2.5:** Crystal structure of the austenitic phase and the tetragonal martensite variants in $Ni_2MnGa$. Arrows indicate the possible orientations of the saturation magnetization $M^{sat}$ within each variant.

The changes in the martensite microstructure of MSM alloy are reflecting in macroscopic magnetic behavior of the material. For example, the net macroscopic magnetization and magnetization curve of the alloy change when the magnetic martensite microstructure is rearranged by a mechanical stress [36]. The onset of MSM effect is observed as a jump of the magnetization in the first quadrant of the M-H curve. This behavior is illustrated in Fig. 2.6 [37].

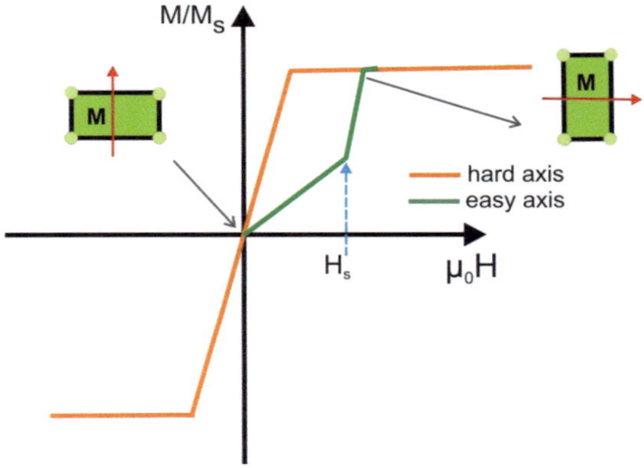


**Figure 2.6:** Schematic illustration of the magnetization characteristics of an MSM sample along both easy and hard axis orientation. The sample is presumed to be in single variant state and magnetization direction in a unit cell is shown at two different position measured along easy axis direction. $H_s$ is the switching filed at which reorientation of martensite varaints start.

Typically, the threshold field ($H_s$) is of the order of 0.05 to 0.2 T [16]. In the reverse direction, the magnetization follows the path determined by the easy axis of the samples. The magnetocrystalline anisotropy ($K_u$) is one of the most important parameters that determine the occurrence of the MSM effect. The high value of $K_u$ represents strong coupling of magnetic and structural order. This coupling generates an interaction between magnetic and crystallographic twins. The anisotropy constants can be obtained by measuring the magnetization curves for different orientations of the field with respect to the crystallographic axes of the material [38]. In tetragonal Ni-Mn-Ga, the easy axis of magnetization points either in the (100) or the (001) direction.

## 2.2 Mechanism for magnetic field induced strain

The giant magnetic field induced strain achievable in Ni-Mn-Ga alloys is due to magnetic field-induced twin-boundary motion. To observe this, the material must exhibit the following characteristics.

- The material should exhibit modulated martensite crystal structure revealing structural order in the material. Thus, the stress required to move twin boundaries must be low. It is illustrated in Fig. 2.7

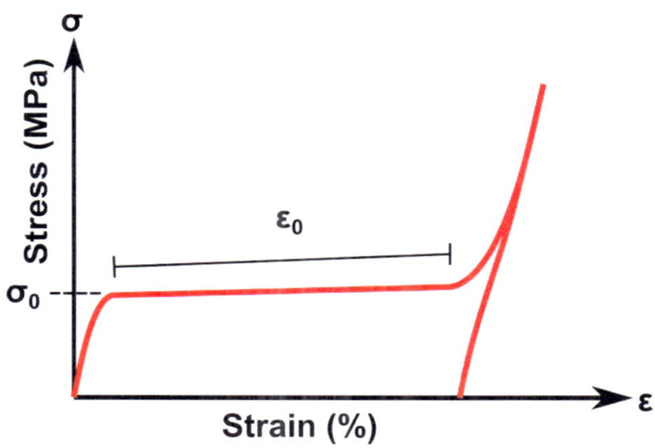


**Figure 2.7:** Schematic of stress-strain diagram for Ni-Mn-Ga under a mechanical stress. The flat portion of the curve corresponds to the deformation thorough twin boundary motion. $\sigma_o$ is the twinning stress and $\varepsilon_o$ is the twinning strain.

- The MSM material should be in ferromagnetic state while in operating environment. Hence the transition from the paramagnetic to ferromagnetic state must be above its martensitic transition.

- The material should exhibit strong magnetocrystalline anisotropy ($K_u$). The reorientation of the magnetic moment must be strongly linked to a particular crystallographic direction (the easy axis). $K_u$ is a measure of the energy required to rotate the magnetization to align with a field directed away from the easy axis.

These requirements can be summarized by the following mathematical relations:

$$K_u > M_s H > \sigma_{ex}\varepsilon_0 > \sigma_0\varepsilon_0 \quad (0.1)$$

Where $K_u$ is the anisotropy energy, $M_sH$ is a measure of magnetic energy input through the applied field ($M_s$ is the saturation magnetization and H is the applied field), $\sigma_{ex}\varepsilon_o$ is a measure of the energy necessary to move twin boundaries ($\sigma_o$ is the twinning stress and $\varepsilon_o$ is the twining strain as shown in Fig. 2.7)

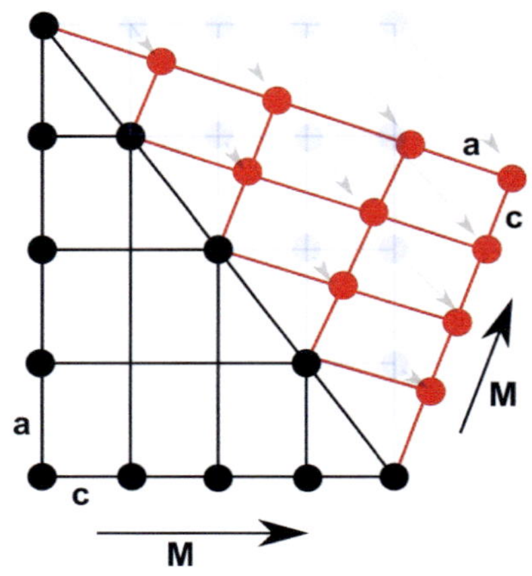


**Figure 2.8:** A two-dimensional schematic diagram of twin boundary mobility in a MSM material with a tetragonal unit cell. The a-and c-axes in each variant are indexed along with the direction of magnetization. The light blue atoms represent the starting positions and the arrows indicate the necessary shear to produce the second variants.

To deform the martensite material, the preferred mechanism in Ni-Mn-Ga alloys is thorough the formation and propagation of twins [39]. A schematic representation of twin boundary mobility with an applied magnetic field direction is shown in Fig. 2.8 for a tetragonal unit cell. The characteristic mirror symmetry at the twin boundary can be seen as well as the change in direction of the short c-axis and the long a-axis. The different orientations of the martensitic unit cell are referred as twin variants. Due to the large value of magnetocrystalline anisotropy ($K_u$) in Ni-Mn-Ga alloys, the magnetic moment aligns with the shorter c-axis in the absence of an applied magnetic field direction. The net magnetization vector (M) aligned with the c-axis and its directions across the twin boundary are indexed in Fig. 2.8.

## 2.3 Ni-Mn-Ga crystal structure

A stoichiometric $Ni_2MnGa$ is a Heusler type ($L2_1$) cubic crystal structure with lattice parameter a = 0.5825 nm at room temperature (RT). The crystal structure undergoes martensitic transformation at 200 K and this transformation has been identified by X-ray diffraction as modulated tetragonal martensite with lattice parameters a = b = 0.5925 nm and c = 0.5563 nm [40]. High-resolution neutron diffraction experiments performed on Ni-Mn-Ga alloys indicate that the crystal structure is approximately tetragonal [41]. The large strain associated with the martensitic transformation is accomplished by the formation of martensitic twin variants separated by twin boundaries so that the internal energy is minimized. The different martensitic twin variants have the same primitive cell but different structural orientation (Fig. 2.2) [42]. There are mainly three observed martensite structures in Ni-Mn-Ga alloys close to stoichiometric composition. They are 10M, 14M and NM modulated martensitic structures.

The martensite structure in Ni-Mn-Ga alloys depends on its chemical composition and thermal stability of the different martensitic phases. There seems to be certain order during cooling from parent austenite phase according to which they form martensite phases or change in the intermartensitic transformation [43]. The Non-Modulated (NM) structure is the most stable one out of the three martensite phases [37, 44] and alloys transforming straight to the NM martensite from the parent phase have typically transforming temperatures close or above the Curie point [45]. The 10M structure transforms directly from the parent phase at lower temperatures close to the ambient temperature, while the 14M martensite is formed upon cooling directly from the parent austenite phase only in a narrow temperature range below the Curie point [24]. The transformation to the stable non-modulated martensite can also take place via the intermartensitic reactions such as (parent phase) P→10M→NM or P→10M→14M→NM [10, 46-48]. Chernenko et al [47] showed that the ratio of the transformation enthalpies is 10:1 in the sequence P→10M→NM. Due to the change in the crystal structure, the intermartensitic reactions seems to set the lowest operating temperature for the magnetic shape memory effect [49].

In Ni-Mn-Ga alloys, the crystal structure strongly correlates with its chemical composition and martensite temperature. Hence a correlation between the average number of valence electrons per atom (e/a) and the tetragonal distortion of the cubic lattice is suggested [50]. The lattice parameters are given in the orthorhombic coordinates in which the axes are parallel to the diagonal directions of the cubic coordinates [51]. This type of approach is convenient to describe the crystal lattice modulations as a long period super structure. However, for a simple description of the strain values induced by stress or magnetic field, the cubic parent phase coordinates are considered [52]. The correlation of the lattice parameters of these different coordinate systems given by Pons et al [51] is

$$a = a_{ort}/2 \quad \text{and} \quad c = c_{ort} \tag{0.2}$$

where a and c refer to the lattice constants given in the cubic parent coordinates and $a_{ort}$ and $c_{ort}$ are the parameters in the orthogonal coordinates.

When applying to the cubic coordinate system, the modulated structures 10M and 14M have a lattice parameter ratio c/a < 1 and in the non-modulated (NM) martensite c/a > 1. It has been shown that c/a > 1 takes place at e/a = 7 [10]. Moreover, this critical value is estimated to be 7.61 to 7.62 [53]. In both cases, the observed change has been identified for NM martensite.

### *2.3.1 10M modulated martensite*

Webster et.al [54] discovered for the first time 10M martensite in the stoichiometric $Ni_2MnGa$ alloy. Later, the crystal structure is identified to be modulated by a transverse diaplacive wave along the [110] direction [44]. The crystal structure of the 10M martensite is tetragonal with c/a = 0.93 and c/a <1. Theoritically, the tetragonal distortion (1-c/a) in 10M $Ni_2MnGa$ unit cell calculated to be 0.6 [55]. 10M modulated martensite can be represented as a periodic shuffling of a long-periodic stacking of {110} close-packed planes [51]. This is illustrated by the schematic in Fig. 2.9.

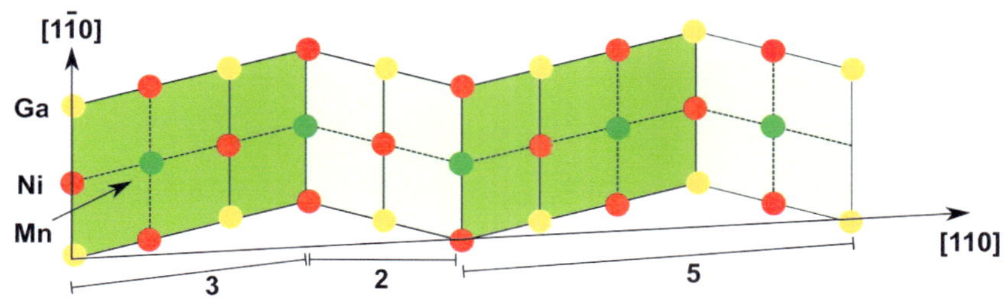


**Figure 2.9:** Schematic representation of 10M modulated crystal structure with a periodic stacking of 10 atomic layers along [110] direction.

The stacking sequence of $(3\bar{2})$ atomic unit cells results in a periodicity of 10 layers. Even though the full symmetry is included in the 10M cell, the structure is referred to as 5M mostly. The modulations can be observed by neutron, electron or X-ray diffraction measurements as four additional spots between the main reflections along the [110]* direction in the reciprocal space.
It has been identified that the tetragonality of the 10M martensitic lattice increases with decreasing temperature and saturates at very low temperatures [55]. At ambient temperature, the magnetic anisotropy of the 10M martensite is in the range $K_u$ =1.45 to 1.67 x $10^5$ J/m$^3$, depending on the alloy composition [56]. The modulation in these types of martensites plays important role for the mobility of twin variants, accordingly largest strains are reported in 10M martensites. Till to date, most of the MSM effect has been investigated in 10M martensite and hence it is often termed the most promising candidate for practical applications.

### *2.3.2 14M martensite*

14M martensite was first discovered by Martynov and Kokorin [44]. The crystal lattice structure is approximately orthorhombic with lattice constants ratio $c/a$ = 0.89 and $c/a$ < 1 with a monoclinic distortion of less than 0.4° [57]. In the orthorhombic coordinates, the lattice of the 14M martensite is seen as monoclinic [51]. 14M martensite is illustrated by the schematic in Fig. 2.10
The 14M martensite was identified in diffraction experiments with a pattern where the distance between the main reflections along [110]* direction in reciprocal space [58]. In 14M martensite the easy axis of magnetization is the short c-axis, while the long a-axis corresponds to magnetic hard axis and the magnetization of b-axis is intermediate one. Consequently, two magnetic anisotropy constants $K_b$ and $K_a$ are needed to characterize the orthorhombic crystal structure. However, it has been found that the crystal structure is extremely sensitive to composition.

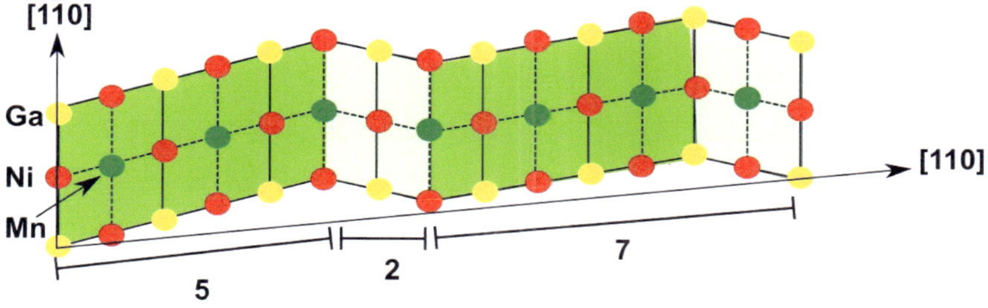


**Figure 2.10:** Schematic representation of modulated orthorhombic structure with a period of 14M layers. The stacking sequence of $(5\bar{2})$ atomic layers results in a periodicity of 14. Even though the full structure is included in 14M, the structure is referred to as 7M mostly.

### *2.3.3 Non-modulated martensite*

The non-modulated (NM) martensite of the Ni-Mn-Ga system is also found by Martynov and Kokorin [44]. The crystal structure is tetragonal with $c/a > 1$, where c is the long axis. It has been identified that the change from $c/a < 1$ to $c/a > 1$ takes place when $e/a > 7$ [59]. As indicated by the acronym the non-modulated martensite does not possess any periodic stacking of atomic layers along [110] direction. The NM phase is the most stable martensitic phase that exist above or far below $T_c$ [44]. But in contrast to 10M or 14M martensite structure, the long c-axis is the magnetic hard axis and the short a- and b-axis are the magnetic easy axes [43]. The non-modulated tetragonal structure seems to possess the best chemical properties among the Ni-Mn-Ga alloy martensites [60]. Its corrosion resistance in artificial seawater was found to be higher than the 14M and 10M martensites.

## 2.4 FSMA material applications

Shape-memory alloys are suitable candidates for actuator and sensor applications due to their large strains and high mechanical durability [20]. It is reported that NiTi and NiTiCu alloys can operate with 2% strain even after millions of cooling-heating cycles. However, the frequency is limited to ≈ 100 Hz due to the thermal nature of the phase-transformation process. The frequency of actuation is ultimately depends very much on the size, surface to volume ratio and thus on the cooling efficiency [3]. Thus, FSMA actuators are attractive for high frequency applications. Recently it has been shown that FSMAs can operate up to the kHz regime for different applications [61]. Moreover, actuators based on FSMAs have a fast response time (less than 0.2 ms) and are able to generate large thrusts [62]. The clear advantages of FSMAs are their ability to generate very large strains and their high energy density at magnetic fields below 1T. In addition, the maximum work output at weaker fields is of the same order of magnitude as that of a standard giant magnetostrictive alloys (0.01 $J/cm^3$ at 0.05 T for TbDyFe) [63]. However, the work output for the best NiTi alloys of almost 10 $J/cm^3$ is noticeable. Moreover, the large strain of 10 % exhibited by FSMA materials due to magnetic field induced strain, magnetostrictive materials are still alternatives to FSMA materials because of the large force they are able to produce [5].

FSMA actuators have been identified to be potential energy harvesters due to good power output from low frequency vibrations, low impendence and high power density [11]. It has been identified that ferromagnetic Heusler alloys ($Co_2MnGe$, NiMnSb) are good candidates for efficient electrical spin injection into (p-i-n $Al_{0.1}Ga_{0.9}As/GaAs$ light emitting diode) semiconductors. Because these materials are predicted to be half metallic and have a band structure with only one occupied set of spin states at the Fermi level resulting 100% spin polarization [64]. In addition most of the Heusler alloys have high Curie temperature ($T_c$ > 600ºC) along with large magnetic moments (> 3.5$\mu_B$ per unit magnetic moments of Heusler alloy compound) [65]. Hence it is interesting to note that these materials can be used as ferromagnetic semiconductors.

## 2.5 Ni-Mn-Ga thin films

First actuation concepts for bulk single crystals [66] have been realized but a broad application is hampered by the high material cost. The advantages of MSM actuation emerge when scaling down to the micrometer range. In this realm, the mechanical properties of most competing actuation principles show a minor down-scaling behavior [3]. For example, the work density of electromagnetic actuators down-scales proportional to $r^2$ compared to $r^0$ for MSM actuators, which do not require an integrated magnetic field source. The actuator design can therefore be kept quite simple without any need of driving electrical contacts compared to microsystems based on piezocrystals. In order to use the MSM effect for microactuation, thin MSM films are thus potential building blocks for robotics as well as for novel actuators and sensors. Furthermore different functional properties such as the phase transformation from austenite to martensite and magnetically induced reorientation can be used simultaneously [67], which increases the application potential even further.
The preparation of thin Ni-Mn-Ga films or foils is achieved in various ways. In the top-down approach one use a Ni-Mn-Ga single crystal as starting material to prepare foils based on thickness reduction process techniques. An alternative route is the bottom-up approach based on sputter deposition. The film deposition on a substrate can be realized by a number of techniques such as molecular beam epitaxy (MBE) [68], pulsed laser deposition (PLD) [69] or sputtering [70]. The resulting film structure depends on various parameters e.g. preparation route, temperature, annealing procedures and substrates. The prepared films can therefore exhibit polycrystalline structure [71], show fiber textures [72] or an

epitaxial relation to the substrate [73, 74]. Since epitaxial films are the thin film counterpart to bulk single crystals, they are most interesting for the application in microsystems. Epitaxial Ni-Mn-Ga films have been prepared by DC magnetron sputtering. Sputtering is the most favorable deposition method to obtain thick films on a large scale as required for microtechnology. Film growth is the first decisive step towards active films with high twin boundary mobility. The deposition parameters such as temperature or working pressure have to be chosen carefully in order to tune the film properties. In addition, the substrate has to fulfill main requirements such that Ni-Mn-Ga would crystallize in the desired phase.

- No chemical reactions should takes place between the film and the substrate
- The substrate surface should be smooth and will be in a single crystalline state to promote crystalline growth
- The thermal expansion coefficients difference between substrate and the film should be as minimum as possible

Semiconductor materials like Si, Ge and GaAs match to the above-mentioned criteria and are thus proper substrate materials for the preparation of Ni-Mn-Ga films. However at high temperature Si and GaAs tend to react with Ni-Mn-Ga and diffuse into Si to form silicide compounds. For these reasons a chemically inert, structurally and thermodynamically compatible sacrificial layers with a thickness of only a few atomic monolayers have been used between the substrate and the film when depositing Ni-Mn-Ga on GaAs by MBE technique. In addition these intermediate layers acts as templates in controlling the film growth and its crystal structure. Possible sacrificial layers of $Sc_{0.3}Er_{0.7}As$ [75], Ni-Ga [76], and ErAs [77] has been used. Semiconductor substrates are not flexible to induce the martensitic transformations and to observe magnetic field induced strain. Hence the film has to be totally or partially released from its substrate.

In addition, epitaxial growth of Ni-Mn-Ga films has been realized on various substrates e.g. GaAs (001) [73, 75], $Al_2O_3$ [71], $SrTiO_3$ [78], and MgO(100) [79-83]. The large variety of suitable substrates for epitaxial growth indicates an easy adaption of Ni-Mn-Ga. In this thesis most of the Ni-Mn-Ga films are deposited on MgO (100) single crystalline substrates. The misfit between MgO and the austenitic Ni-Mn-Ga is -27.6% if one assumes a cubic-on-cubic growth. This misfit is reduced to 2.4% once the Ni-Mn-Ga unit cell is rotated by 45° with respect to the substrate edges. The growth direction of Ni-Mn-Ga is determined in the austenitic temperature regime. During cooling, the film transforms to the martensitic phase. The habit plane between austenite and martensite is close to a (101) type plane [80]. Since the austenite orientation is fixed by the substrate, the tetragonal and orthorhombic martensites have to adopt this orientation. Hence in Ni-Mn-Ga films the formation of modulated martensites are considered to be adaptive phases, which was explained through the concept of adaptive martensite by Kaufmann et al. [84].

### *2.5.1 Concept of adaptive martensite*

Giant strains are achieved by a rearrangement in twinned microstructures only in a materials exhibiting modulated crystal structure. The modulated crystal structures in bulk Ni-Mn-Ga alloys are considered as thermodynamically stable phases. These modulated phases form by a displacive transition from a high symmetry austenite to a low-symmetry phase, which is of the martensite type as shown in Fig. 2.11. The transformation requires that this martensite is accommodated to a habit plane as a lattice invariant interface which fixes the geometrical relationship between the two crystal structures [85]. The lattice mismatch is compensated as twinning of martensite. Hence large number of twin boundaries connecting differently aligned martensite variants are introduced. Extrapolating this geometrical continuum approach to atomic scale, modulated structures observed in materials with lattice instabilities can be understood as ultrafinely twinned metastable structures and not as a thermodynamically stable phase. In this view, the large and complex apparent unit cell of the modulated phase is composed of nano-twin lamellae of a simpler, thermodynamically stable, martensitic phase.

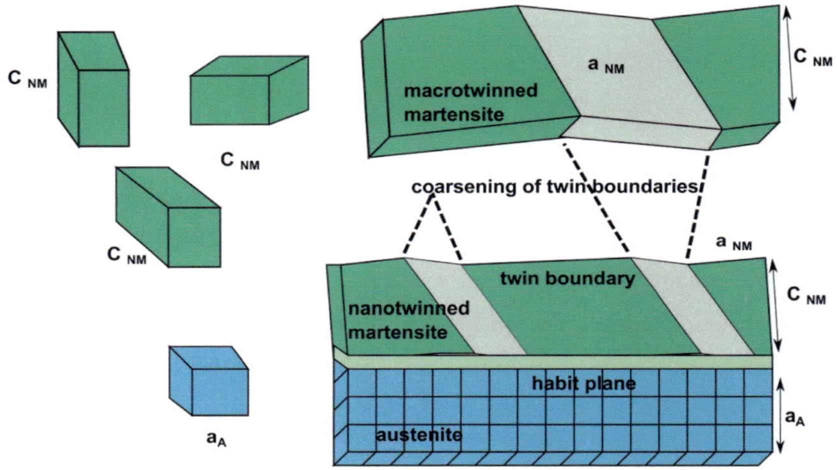


**Figure 2.11:** Schematic of the martensite microstructure formed on a cubic austenite crystal phase. Left: sketch of the orientation relationship between parent austenite and nanotwinned (adaptive) martensite phase. The different green background colors mark differently orientated tetragonal martensitic variants, which are connected by twin boundaries. The light green plane marks the habit plane. A macrotwinned martensite where the twin boundaries have a macroscopic distance compared with the atomic distances shown in nanotwinned martensite.

The twinning periodicity and, hence the modulation is determined by geometrical constraints and the transformation path. The key requirement for the validity of this explanation is of very low nanotwin boundary energy. The suitability of this concept for the Ni-Mn-Ga system can be understood by explaining the sequence of intermartensitic phase transitions [86] and stress induced martensite [87]. The concept of adaptive martensite competes with alternative theoretical ideas, e.g, the stabilizing effect of Fermi surface nesting for modulated phases in metallic martensites [51]. These concepts are suggested as an expansion of the well-accepted theoretical and experimental observation of an electron instability [41].

The experimental proof for the adaptive nature of modulated martensite structures is difficult because the thermodynamic measurements do not easily identify metastable phases. Furthermore, diffraction experiments cannot directly distinguish a regular nanotwinned microstructure from a long-period modulated phase [88]. It has been identified that Ni-Mn-Ga thin films are suitable candidates to decide on the nature of modulated phases because of two main reasons. First one, the geometrical constraint at the interface with the rigid substrate stabilizes otherwise thermodynamically unstable phases and makes these intermediate states accessible to experiments. Second, the single crystalline substrate acts as a geometrical reference frame, which allows determining crystallographic orientations of all phases in absolute coordinates.

# Chapter 3

# Experimental methods and characterizing techniques

This chapter explains the various experimental techniques used in this thesis work. A brief introduction to foil preparation (top-down approach) and thin film deposition on single crystalline substrate (bottom-up approach) is made. Further, material characterization, training, mechanical and magnetostrain characterization techniques are discussed in detail. Finally, Digital Image Correlation (DIC) method adapting to thin freestanding films is introduced and explained.

## 3.1 Preparation of Ni-Mn-Ga foils

Two different kinds of preparation methods are being used to develop Ni-Mn-Ga foils from a bulk reference Ni-Mn-Ga alloy. The methods are distinguished according to the successive process steps involved to fabricate Ni-Mn-Ga foil. They are (a) electric discharge machining and (b) abrasive wire sawing. Bulk Ni-Mn-Ga alloys for the top-down approach are selected by their chemical composition and method of preparation.

For electric discharge machining, the bulk Ni-Mn-Ga alloy single crystals are grown by a new crystal growth technique at Han Meitner Institute, Berlin. SLARE (SLAg Refinement and Encapsulation) is a proprietary modified Bridgeman technique for single crystal growth especially for the metallic alloys, which form highly reactive melts or contain elements with a high vapor pressure like manganese (SLARE technique developed at Hahn Meitner Institute, Berlin). During the growth process an alloy-dependent slag formation occurs, which can encapsulate the metallic melt during the growth process [89]. Thus, it is possible to avoid the loss of manganese and the formation of new seed crystals and as a result the probability of growing a single crystal of high quality with pre-weighed composition increases. Furthermore, slag removes most of the impurities such as oxides and sulfides from the metallic melt during the growth process. SLARE enables the growth of single crystalline rods of 80 mm length and 16 mm diameter with a mass of about 110 grams. Single crystalline Ni-Mn-Ga foils are also prepared from a $Ni_{49}Mn_{31}Ga_{20}$ (at.%) bulk single crystal at Han Mettner Institute, Berlin. Compared with the Bridgman growth method, this new technique allows producing a single crystal with low impurities and with extremely small mosaic spread.

To prepare foils by abrasive wire sawing, a Ni-Mn-Ga bulk single crystal grown by AdaptaMat Ltd., Finland was used as a reference sample. In addition, the Ni-Mn-Ga foils are also prepared at same place. The chemical composition of bulk Ni-Mn-Ga reference alloy is determined by energy dispersive X-ray (EDX) analysis to be about $Ni_{50.2}Mn_{28.4}Ga_{21.4}$ (at.%). Bulk Ni-Mn-Ga alloy ingots were produced using a modified Bridgman method.

### *3.1.1 Electric discharge machining*

Plates of 1 mm thickness are initially cut from a bulk single crystal $Ni_{49}Mn_{31}Ga_{20}$ along (100) planes using electrical discharge machining (EDM) followed by electro polishing in $CH_3OH/HNO_3$ solution. The schematic shown in Fig. 3.1(a) illustrates the successive steps involved while preparing the Ni-Mn-Ga foils. Wet-mechanical abrading and mechanical polishing are then used to obtain the desirable foil thickness. Finally, the foils are chemo-mechanically polished using $SiO_2$ in an acid aqueous dispersion. Based on this technique foils having 200, 80, 60 and 30 μm are prepared after successive preparation steps. All these foils are investigated to study the influence of their surface damage on the crystal structure, phase transformation, mechanical properties and finally magnetic field induced reorientation in Chapter-5 (section 5.1).

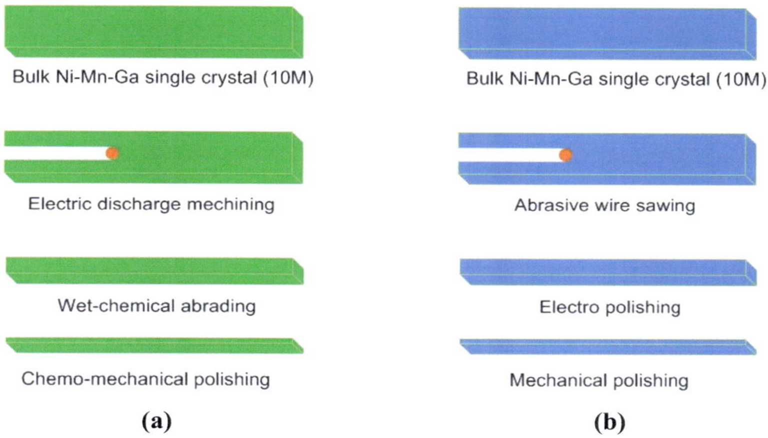


**Figure 3.1:** Schematics of the successive foil preparation methods following by (a) electric discharge machining (b) abrasive wire sawing.

### *3.1.2 Abrasive wire sawing*

The Ni-Mn-Ga foils are cut from a single variant bulk Ni-Mn-Ga alloy by abrasive wire sawing technique at ambient temperature. Since the foil preparation methods involve mechanical treatments, the deformed surface layer due to mechanical treatment is removed by electropolishing. Fig. 3.1(b) shows a schematic of the successive steps involved to prepare Ni-Mn-Ga foils by abrasive wire sawing. Desired thickness of the foils is achieved by using wet-mechanical abrading and polishing. By this approach, foils of various thicknesses are adjusted to be about 200 μm and least minimum thickness of 30 μm. However, finally Ni-Mn-Ga foils of 200 and 30 μm are investigated to study their actuation in magnetic field in Chapter-5 (section 5.2)

## 3.2 Material characterization

### *3.2.1 X-ray diffraction*

The structure of the freestanding epitaxial Ni-Mn-Ga films is analyzed by X-ray diffraction at room temperature. The experiments on as-released freestanding films are performed in reflection mode and for strained freestanding films in transmission mode at the ANKA (Angströmquelle Karlsruhe) synchrotron source. The cross-section of the beam is 1 x 2 $mm^2$, the divergence is 0.05 mrad. The measurements are carried out using a scintillation detector with a Soller collimator. The $2\theta$ angle is varied in the range of 13° and 53°. In addition to the conventional X-ray diffraction, temperature-dependent X-ray diffraction measurements are carried out. The temperature is varied between 25 and 200 °C. The measurements are performed on a freestanding Ni-Mn-Ga film. The tilt angles are varied in between 1 to 10 degrees in steps of 1 degree. To observe all differently orientated lattice planes at different tilt angles, the final XRD diffractogram is generated by adding the $\theta$-$2\theta$ scans for all individual tilt angles for a constant rotation angle.

### *3.2.2 Transmission Electron Microscopy (TEM)*

TEM experiments are performed using a FEI TITAN 30-800 operated at 300 kV. The information limit in the TEM mode is 0.08 nm. A nominal spot size of 0.14 nm is used for high angle annular dark-field (HAADF) imaging in scanning transmission electron microscopy (STEM). Sample preparation for TEM experiment includes several successive processing steps. The freestanding Ni-Mn-Ga film is fixed to a rigid support at both ends and then a small region of the sample is cut along the cross-section with Ga+ ions (FIB FEISTRATA 400 STEDM). Final polishing with Ga+ ions is performed at 5 kV to smooth the surfaces without having mechanical inhomogeneities. Ultimately, Ga+ ion polishing is performed to reduce the thickness of the lamella to about 30 nm. Fig. 3.2 shows the Ni-Mn-Ga TEM lamella after polishing. At this small thickness levels, the Ni-Mn-Ga lamella is fixed to a copper-handling probe using platinum deposition. At this thickness levels the sample is ready to use for TEM experiments

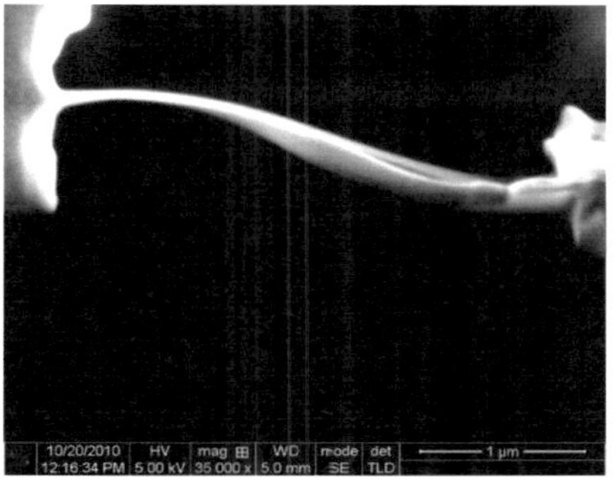


**Figure 3.2:** The final 30 nm thick Ni-Mn-Ga TEM lamella after successive preparation methods by Focused Ion Beam (FIB).

## 3.3 Thermally induced phase transformations

The phase transformation characteristics of the Ni-Mn-Ga foils and thin films are investigated with two different characterization techniques. One is Differential Scanning Calorimetry (DSC), the second one is four-probe resistance measurement. In this thesis, Ni-Mn-Ga foils are investigated by DSC, while the Ni-Mn-Ga films are only investigated by four-probe resistance measurements because of the thin film geometry.

### *3.3.1 Differential Scanning Calorimetry (DSC)*

DSC is a thermo analytical technique in which the difference in the amount of heat required to increase the temperature of a sample and reference is measured as a function of temperature. In order to probe the phase transformation temperatures of shape memory alloys, the DSC experiment is used. In this experiment, the samples to be investigated are kept in small aluminum crucibles with an inner diameter of 5 mm. Sample and reference sample are subjected to a constant heating and cooling rate of 10 K/min between -100ºC and 150ºC temperature cycles.

Initially the heat flow is increased to heat the sample to high temperatures and then decreased to cool down the sample temperature. The DSC curve is characterized by observing an exothermic peak on cooling corresponding to the phase transformation of martensite ($M_p$) and an endothermic peak during heating corresponding to the phase transformation of austenite ($A_p$), respectively.

### *3.3.2 Temperature-dependent resistance measurements*

Resistance measurements are performed using four probing wires at two point contacts on the shape memory sample. The probing contacts were made on a Ni-conducting sheet, which is further contacted to the sample by using a gap-welding machine. The experiments are performed in a thermostat allowing precise control of ambient temperature by maintaining high vacuum inside the cryostat chamber. The temperatures are ramped stepwise through the phase transformation region of the sample while the corresponding electrical resistance is monitored. The base pressure of the thermostat is maintained under high vacuum to avoid oxidation. Depending on heating or cooling direction, the corresponding jump in the resistance is monitored and identified as martensite to austenite or austenite to martensite phase transformation, respectively.

## 3.4 Thermo-mechanical training

Bulk specimens are usually subject to initial compressive training in order to align the martensitic variants and to improve the mechanical performance.

Compression of foils or thin films in lateral direction, however, is not appropriate as buckling effects are likely to occur. Thus, compressive loading may only be applied perpendicular to the foil surface, which appears to be increasingly problematic for decreasing thickness as large contact forces inhibit variant reorientation. Therefore alternative ways of training the foils have been developed taking into account their quasi-two-dimensional constraints.

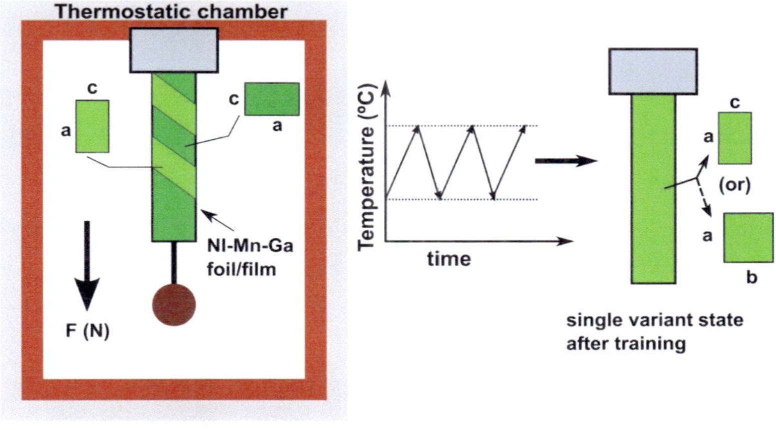


(a) (b) (c)

**Figure 3.3:** (a) Schematic of the thermo-mechanical training procedure for Ni-Mn-Ga foils/films under tensile training in thermostatic chamber, while the sample is being subjected to constant load. (b) Illustration of temperature cycles (c) The final stage of the sample after training.

Tensile load is applied along the long side of the Ni-Mn-Ga specimen close the austenite finish temperature at a stress level above the stress plateau. Then, the foil is subjected to a thermal cycling procedure through the phase transformation temperature regime. Ni-Mn-Ga specimen presumes to be exhibiting 10M tetragonal crystal structure with long a- (=b) axis and short c-axis. Initially, Ni-

Mn-Ga film exhibits multivariant state with a-axis and c-axis are being orientated randomly as shown in Fig. 3.3 (a). This training method allows the preferential alignment of one of the long a-axis along the tensile direction while cooling from austenite. Thus, after thermo-mechanical training the orientation of short c-axis will be either in lateral direction or out-of-plane direction (Fig. 3.3(c)).

## 3.5 Mechanical experiments

The mechanical properties of freestanding Ni-Mn-Ga films are investigated using INSTRON micro tensile testing machine. The experiments are performed in strain-controlled mode. The sample preparation for the tensile test is a crucial step because buckling effects are likely to occur due to film handling and bonding to rigid support, which can induce cracks at the sample edges.

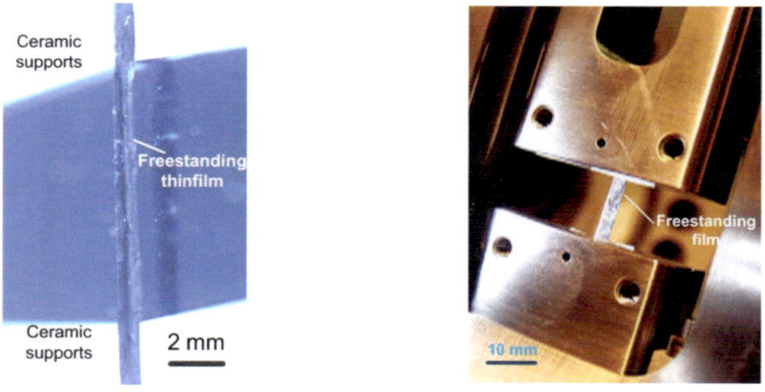


(a) (b)

**Figure 3.4:** (a) Bridge shaped Ni-Mn-Ga specimen before the tensile experiment. (b) Experimental tensile test setup, which is operated in strain-controlled mode.

Freestanding thin films having a width of 1 mm and thickness of 2 µm are fixed to the ceramic support by adhesive bonding on both ends of the film. The final bridge shaped freestanding Ni-Mn-Ga film is shown in Fig. 3.4(a). Ni-Mn-Ga test stripes are strained by applying a force at one end, which is monitored using a 10N load cell. The experiment is controlled using LABVIEW software. The tensile test setup is shown in Fig. 3.4(b).

## 3.6 Magneto-strain experiments

Magneto-strain experiments have been performed in GMW ELECTROMAGNET with the pole diameter of 150 mm (6 inch). The pole gap can be adjusted between 0-127 mm, within 3 mm the maximum field is above 2T. The strength of the magnetic field is controlled by an external power supply. It is measured by using an external Gauss meter with spatial resolution of 3 mT. An easy adjustable nonmagnetic aluminum table is designed and fabricated allowing to move the sample position freely within the magnetic poles. The micro-scale deformations of the Ni-Mn-Ga samples are monitored with a KEYENCE VM-6000E optical microscope.

## 3.7 Digital image correlation

Digital image correlation (DIC) techniques are becoming popular, especially in micro-and nanoscale mechanical testing applications due to their relative ease of implementation and use. Advances in digital imaging and increasing computational resources have been enabling technology for this method. While white-light optics has been the prodominent approach, DIC has recently been extended to SEM/FIB and AFM. Beyond the ability of image-based methods to provide a platform to the events that are occurring during deformation, these techniques were applied to the testing of freestanding thin films or foils because these offer a full field description and are relatively robust at tracking a wide range of 'markers' and varying surface contrast.

DIC has proven to be very effective at mapping deformation in macroscopic mechanical testing, where the application of specular markers (e.g. paint, toner-power) or surface finishes from machining and polishing provides the needed contrast to correlate images well. However, these methods for applying surface contrast do not extend to the application of freestanding thin foils because of mirror-finish quality of surface. Typically this challenge can be solved by applying paint that results in a random speckle pattern on the surface, although the large and turbulent forces resulting from either spraying or applying paint to the surface of a foil are too high and would break the specimens.

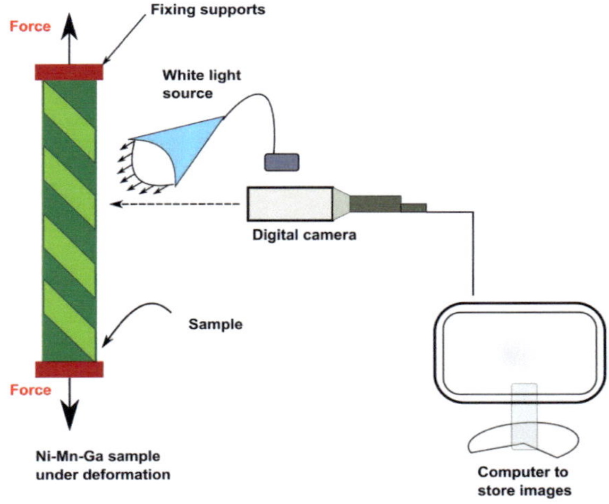


**Figure 3.5:** Schematic of the digital image correlation, where foil was subjected to tensile deformation stress.

A different approach was employed in this thesis work, namely the application of fine powder particles of $SiO_2$ having a diameter of 5 μm that electrostatically adheres to the surface of the foil and can be digitally tracked. This new approach is especially advantageous for metallic foils, which exhibit optically reflecting surfaces. Light blanket of powder particles coat at the gage section of the tensile sample and the larger particles have to be blown away gently to get a clear surface contrast. The remaining particles are those with the best adhesion to the surface, and under low-angle grazing illumination conditions. While the surface contrast present is not ideal for DIC, the high intensity ratio between the particles and the background provide a unique opportunity to track the particles between consecutive digital images taken during deformations. This can be achieved quite straightforwardly using digital image processing techniques.

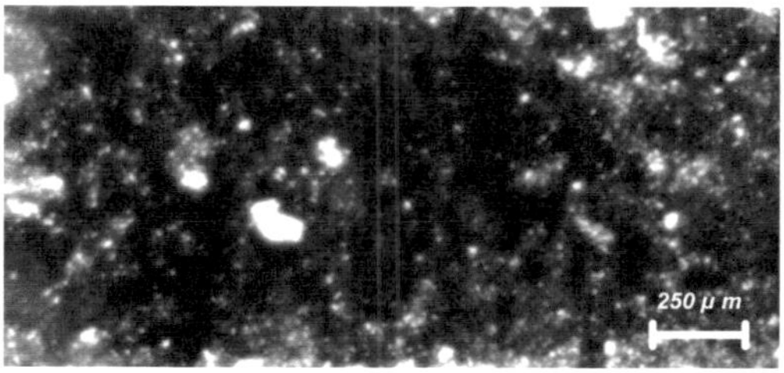


**Figure 3.6:** Representative image of the microspecimen gage section with fine silica particles for digital tracking in strain measurement.

The schematic of the experimental set-up is shown in Fig. 3.5, where Ni-Mn-Ga foils are subjected to uniaxial loading in a tensile testing machine. Simultaneously, during deformation the surface image is captured with a high-resolution camera. The image of such Ni-Mn-Ga foil is shown in Fig. 3.6. Series of images are taken at different deformation levels. All successive images during deformation are analyzed using MATLAB program [90].

To attain tracking with subpixel resolution, a novel image-based tracking algorithm and MATLAB code developed by Rob Thompson and Dan Gianola [91] was used to calculate the strain field in the present investigation. The image pixel resolution that one can achieve in practice using these image-based techniques depends on a number of factors, including lens optical quality, $SiO_2$ particle size and quality, but do not depend on the optical resolution of the image. The $SiO_2$ particles adhere to the sample surface due to Van der Walls force between sample surface and particle, so larger the particle size results in detach of the particles from the surface. In this thesis work, DIC is being used to calculate the strain field distribution during deformation of Ni-Mn-Ga films or foils. The results are presented in chapter 5.

# Chapter 4

# Fabrication technology and design layouts

To enable actuation in a Ni-Mn-Ga film by variant reorientation, freestanding films are required to overcome the constraint by a rigid substrate. The repeated elongation and contraction of Ni-Mn-Ga film are hindered on a rigid substrate such as MgO. The elastic modulus of MgO is in the range of GPa and thus exceeds the blocking stress for magnetically induced reorientation which is only a few MPa. Thin soft polymers can be used as a substrate material, which can exhibit low E-modulus values, but thin film deposition can be a big challenge. Instead, it is worth to consider releasing the thin films from the substrate either by using sacrificial substrate technology or by sacrificial layer technology (chapter-6). This chapter discusses the sacrificial substrate technology and presents a process flow to release the Ni-Mn-Ga films from the substrate material. The basic idea is to prepare the Ni-Mn-Ga films on a substrate, which can be dissolved after deposition. NaCl substrate fulfills this requirement, since it can be easily dissolved in water. Hence, in order to release the Ni-Mn-Ga films from NaCl substrate and for further integration, sacrificial substrate technology is being developed. Finally, different actuation schemes are discussed towards realization of microactuators based on Ni-Mn-Ga freestanding films by considering different boundary conditions (e.g. fixation, tensile loading).

## 4.1 Sacrificial substrate technology

The Ni-Mn-Ga films are grown on a NaCl (100) substrate by DC magnetron sputtering at 480°C. The target has been polished from an arc-melted disc of the composition $Ni_{49}Mn_{30.5}Ga_{20.5}$ (at.%). In this case, typical deposition parameters have been 0.05 mbar argon pressure and 100 mA current at 320 V, resulting in a deposition rate of 1.3 nm/s at a substrate distance of 30 mm. The chemical composition of the films is determined by EDX to be $Ni_{52.8}Mn_{30.8}Ga_{16.4}$ (at.%). The austenite and martensitic phase transformation temperatures $A_{s/f}$ and $M_{s/f}$ have been determined to be 338/399K and 379/320 K, respectively. The Curie temperature is 375 K. Released films show a strong texture with (110) orientation contrary to freestanding Ni-Mn-Ga films grown in MgO (100) substrates.

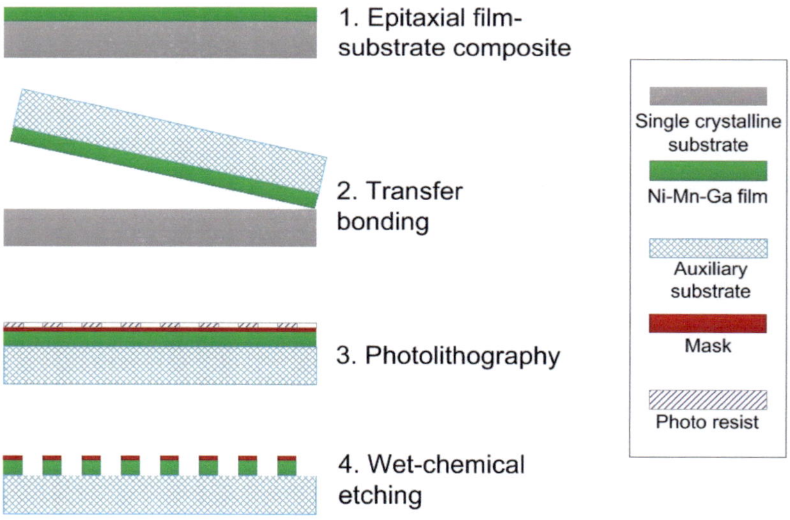


**Figure 4.1:** Process flow for fabrication of epitaxial film actuators based on the sacrificial substrate technology.

After sputter deposition, an auxiliary substrate for the micromachining is bonded to the Ni-Mn-Ga film. Due to the relatively low adhesion between Ni-Mn-Ga film and NaCl substrate, the composite of Ni-Mn-Ga film and auxiliary substrate can be detached without the need to dissolve the NaCl substrate. Subsequently, micromachining is performed by photolithography and wet-chemical etching. Fig. 4.2(a) shows a corresponding Ni-Mn-Ga test microactuator on a micromachined alumina substrate. A typical deflection-temperature characteristic of a Ni-Mn-Ga film microactuator of 1 μm thickness is shown in Fig. 4.2(b) for a constant load of 1 mN upon heating and cooling. In this case, a reversible thermally induced phase transformation occurs between 320 and 400K.

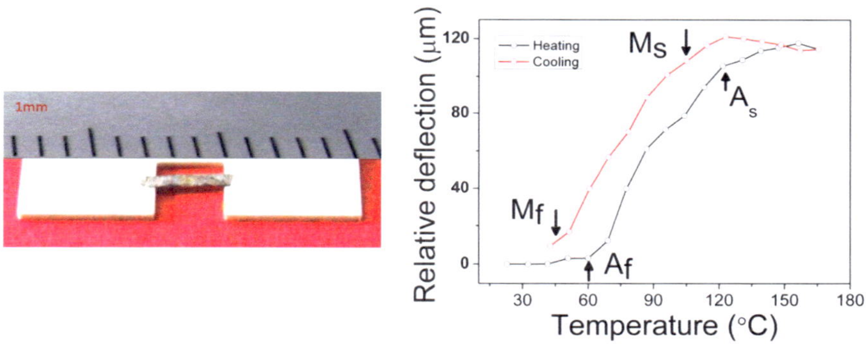


**Figure 4.2:** (a) Test actuator of a freestanding Ni-Mn-Ga film actuator of 1 μm thickness fabricated by sacrificial substrate technology. (b) Deflection-temperature characteristic of the Ni-Mn-Ga film actuator shown in figure (a) for a load of 1 mN.

Despite the easy process steps to release the Ni-Mn-Ga film from NaCl substrate, this technique has certain limitations mainly because of two reasons. First, the lattice mismatch between the NaCl and austenitic Ni-Mn-Ga is -3.1 % for cube-on-cube growth can yield different epitaxial relations for NaCl (100) substrates. Second one is due to hygroscopic nature of NaCl. Before deposition of the Ni-Mn-Ga film, careful bake out of the water content within the substrate is required. Even small amount of water can lead to the complete film

delamination instead of the required continuous film. These limitations are the reason why sacrificial substrate technology is hardly reproducible and not recommended for large-scale applications.
In addition to this above mentioned sacrificial substrate technology, a novel and reproducible sacrificial layer technology has been developed for Ni-Mn-Ga films on MgO substrate with Chromium sacrificial layer. This part will be discussed in chapter 6.

## 4.2 Design layouts

Freestanding Ni-Mn-Ga films or microstructures further need to be integrated in MEMS compatible system with suitable biasing mechanism, which does not require manual adjustments. However it is principally difficult to describe design rules before learning from the experiments. But based on experimental observations on the freestanding Ni-Mn-Ga films and foils three simple actuation mechanism types are distinguished. One type is based on linear actuation and remaining two types are based on out-of-plane actuation. The actuation mechanisms are considered depending on the magnetic field configuration and the methods of stress biasing. The material of interest is a Ni-Mn-Ga film showing a 10M tetragonal martensite structure.

### *4.2.1 Principle of actuation in Ni-Mn-Ga films*

Actuation in Ni-Mn-Ga material is due to magnetic field induced reorientation (MIR) of martensitic variants. Fig. 4.3(a) shows a schematic representation of MIR actuation in bulk Ni-Mn-Ga single crystals. For crystallographic structures showing MIR (10M) the short c-axis coincides with the magnetic easy axis. After transformation from cubic austenite, the martensite crystal consists of a mixture of tetragonal martensite variants having different magnetic and crystallographic orientations. Regions of different c-axis orientation are separated by twin boundaries, which are highly mobile due to the low twinning stress.

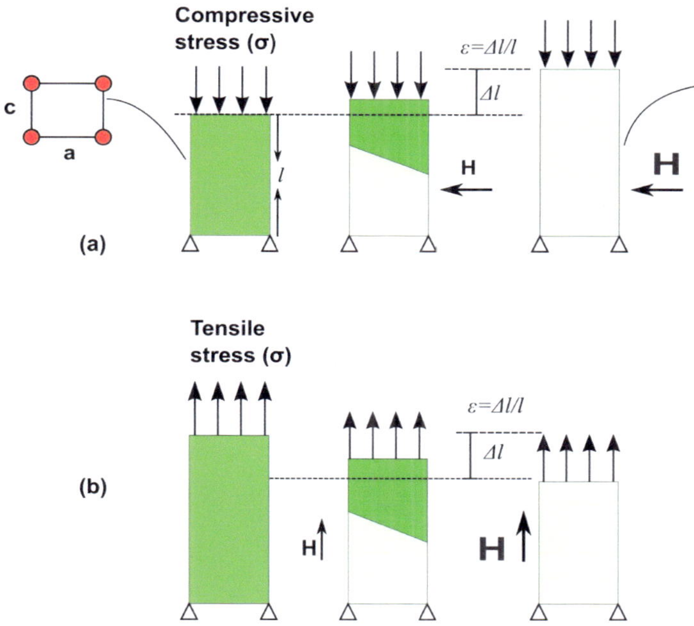


**Figure 4.3:** (a) Principle of MSM actuation in single crystalline bulk Ni-Mn-Ga actuator for the case of compressive loading; (b) principle of MSM actuation in a single crystalline thin film actuator for the case of tensile loading.

In bulk Ni-Mn-Ga single crystal the initial state is usually adjusted by a compressive stress perpendicular to the magnetic field. For thin film specimen, however this kind of preconditioning is not possible as buckling is likely to occur upon compressive loading in the film plane. In this case, an alternative way of preconditioning has to be developed. Fig. 4.3(b) shows the option of tensile loading the film along the long side of the specimen, which favors the orientation of one of both a-axes (10M) being in tensile direction. By applying a magnetic field along the tensile loading direction, a single variant state with final c-axis orientation being parallel to the magnetic field is obtained. In this case, however the shape change is smaller than maximum strain ($\varepsilon_{max}$) as this kind of preconditioning does not result in an initial single variant state.

### *4.2.1 Ni-Mn-Ga film linear actuation*

Linear actuation of Ni-Mn-Ga films can be achieved either by applying counteracting magnetic fields or by applying tensile load counteracting to the magnetic field. Since the basic actuation mechanism in foils or thin films is the same especially in the case of linear actuation, the actuation design and performance characteristics are discussed in detail in chapter 5.

### *4.2.2 Out-of-plane Ni-Mn-Ga film actuation*

The out-of-plane actuation of Ni-Mn-Ga films is being considered because larger deformations can be achieved than a linear actuation (in-plane tensile strain). In addition, it is being possible with simple design by fixing the Ni-Mn-Ga film at two ends of rigid carrier support, while the movable center part is used to generate work. Two different operative mechanisms are proposed to achieve reversible actuation by using MIR. In mechanism I, actuation can be observed by applying a tensile point force acting in out-of-plane direction. The deflection of the bridge shaped actuator is reset to original position by applying a homogeneous magnetic field. The angle 'α' describes the deflection of the Ni-Mn-Ga film from its in-plane direction. In mechanism II, actuation can be observed by applying a tensile magnetic stress in out-of-plane direction and deflection is again reset to original position by applying in-plane homogeneous magnetic field. The two mechanisms are schematically illustrated for a simple bridge shaped actuator in Fig. 4.4.

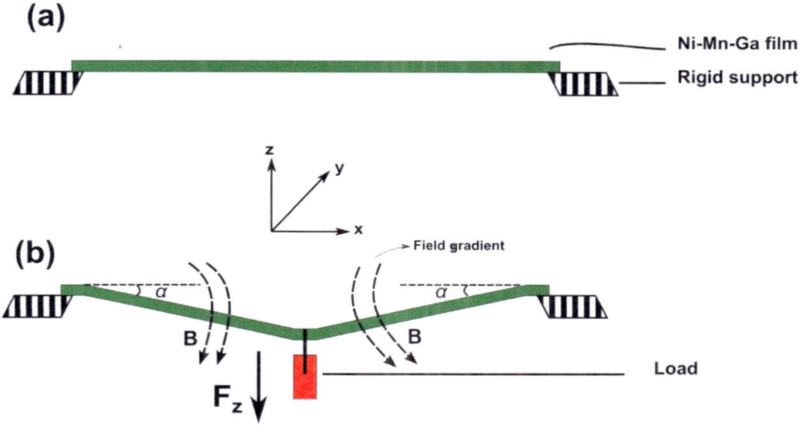


**Figure 4.4:** Schematic of the Ni-Mn-Ga film bridge actuator which is being fixed on both ends (a) without tensile stress (b) with tensile load or magnetic field (B) gradient acting in out of plane direction.

The underlying material consists of 10M martensite with three tetragonal variants having a short c-axis and two long a-axes as shown in Fig. 2.5. In each case, it is assumed that the Ni-Mn-Ga bridge actuator is in an initial single variant state with short c-axis being oriented in x-direction (variant 1).

To develop a microactuator based on MSM actuation, the Ni-Mn-Ga films have to be integrated along with magnetic field coils, loading elements in order to observe the reversible actuation, which is described above by two operative mechanisms. Therefore, integrating a bridge type Ni-Mn-Ga film, mechanical load and magnetic field coils illustrate a microactuator with a reversible actuation mechanism as shown in Fig. 4.5 and Fig. 4.6. The operative function of Ni-Mn-Ga film in MSM microactuator is described as follows:

A point force $F_z$ is applied in z-direction to elongate the actuator as shown in Fig. 4.5(a). The force $F_z$ mainly generates a tensile stress along x-direction that favours variants 2 and 3 with the long a-axis being aligned in x-direction. By applying a magnetic field (B) in x-direction, the initial single variant state can be restored by magnetic field induced reorientation (MIR) as in Fig. 4.5(b). Thus, the initial deflection can be reset depending on the point force $F_z$.

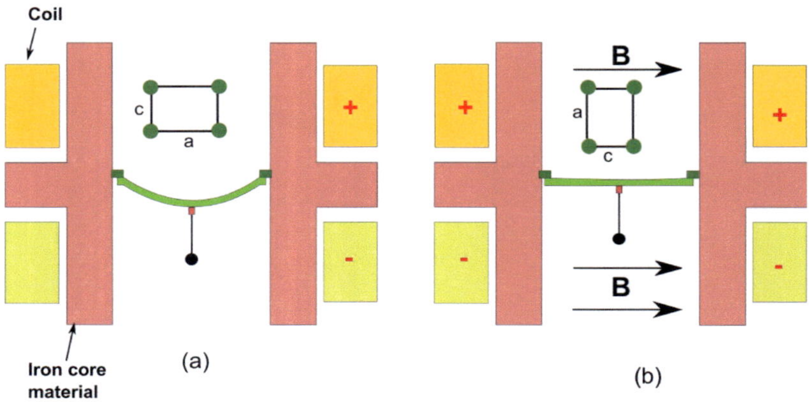


**Figure 4.5:** Schematic of the bridge actuator based on mechanism I, (a) deformation by tensile stress in out-of-plane direction (variant II), (b) original position is reset by magnetic field (B) (variant I).

Alternatively, reversible actuation is possible by the use of two different magnetic field configurations to avoid any mechanical loading on the variant configuration. In this case a magnetic field gradient is applied in x-z plane that acts on the magnetic moments of the Ni-Mn-Ga bridge causing a deflection in out-of-plane direction as shown in Fig. 4.6(a). By applying a homogeneous magnetic field along x-direction, MIR restores the original single variant state as in Fig. 4.6(b). Since the forces for deflection and resetting can be switched on and off sequentially, biasing forces can be avoided during resetting that may hinder complete resetting. In a proper magnet design, the switching between field gradient and homogeneous field can be achieved by just reversing the current direction in one of the electric coils

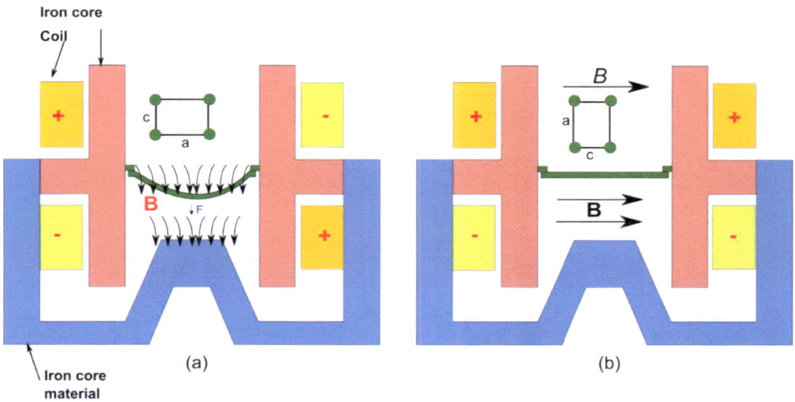


**Figure 4.6:** Schematic of the actuation mechanism under mechanism II, (a) tensile stress by a magnetic field (B) gradient along z-direction (variant 2), (b) reset to original position by using magnetic field (B) along x-direction (variant1).

Three different modes of actuation based on the MSM effect is possible in Ni-Mn-Ga films. To develop such kind of microactuators importantly the Ni-Mn-Ga films should exhibit modulated crystal structure allowing low twinning stress favorable for variant reorientation. In addition to this, the integration elements such as magnetic coils and reset elements are necessary to achieve actuation. However, bridge type of actuation principles are not presented in this thesis because no MIR has been achieved so far in Ni-Mn-Ga thin films.

Here, two different approaches are being followed to develop FSMA linear microactuators based on MIR.

1). The top-down approach of thickness reduction of bulk Ni-Mn-Ga single crystals to foil specimens of decreasing thickness (200-30 μm) and subsequent integration (see chapter-5).

2). The fabrication of free-standing epitaxial Ni-Mn-Ga thin film actuators in a bottom-up manner by magnetron sputtering, substrate release and integration technologies (see chapter-6).

# Chapter 5

# Miniaturized Ni-Mn-Ga foil actuators

In this chapter, investigations are carried out to fabricate and develop miniaturized Ni-Mn-Ga actuators by top-down approach using Ni-Mn-Ga foils. Before studying the performance characteristics, material and mechanical properties are investigated. Since the foils are prepared by using two different fabrication routes, one is by using electric discharge (ED) machining and another one is by abrasive-wire sawing (AS), the material and mechanical properties are entirely different. Finally, the development of miniaturized Ni-Mn-Ga linear actuator will be presented.

## 5.1 Surface effect on twin microstructure

Ni-Mn-Ga foils are prepared by using electric discharge machining of bulk reference Ni-Mn-Ga alloy exhibiting 10M modulated martensite at room temperature (RT) [89]. Since the foil preparation involves severe mechanical processing methods, the surface damage plays an important role in affecting the functional properties induced by temperature and external magnetic field. In order to evaluate this effect, foils of different thicknesses in the range 30 -200 μm are investigated

### *5.1.1 Crystal structure*

The crystal structure of the Ni-Mn-Ga foils with 200 and 30 μm thickness is investigated by using X-ray diffraction (XRD). Depending on the foil thickness, high intense peaks are observed as shown in Fig. 5.1. For the 200 μm foil, a high intense peak occurs at the 2θ value of 62.5 degree in agreement with a (400)

10M reflection as observed in bulk Ni-Mn-Ga single crystals [92, 93]. By decreasing the foil thickness down to 30 μm, the high intensity peak appears at 68 degree, this peak is corresponding to (004) 14M martensite [92].

From XRD measurements, the full width at half width maximum (FWHM) of the XRD peaks is calculated to study the influence of mechanical stress on the foil microstructure. For 200 and 30 μm foils, FWHM is calculated to evaluate the structural change caused by mechanical treatment. It reveals that for the particular 2θ value, FWHM increases from 0.08 to 0.17 degrees with decreasing foil thickness from 200 to 30 μm. Chmielus et al [92, 94] investigated extensively the surface damage of Ni-Mn-Ga alloys on the twin microstructure, where they found out similar features from XRD measurements

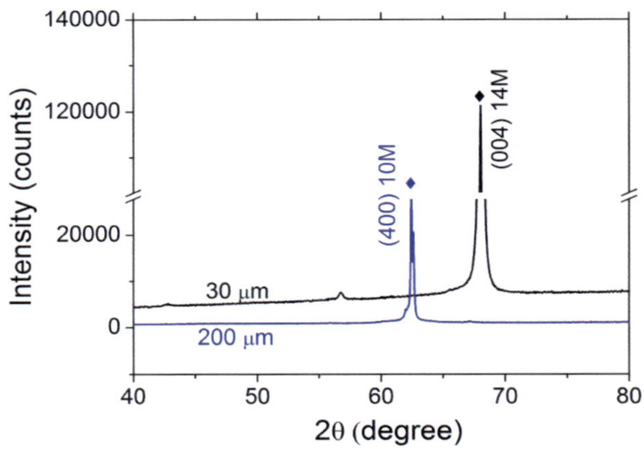


**Figure 5.1:** X-ray diffraction characteristics performed in Bragg-Brentano geometry by using Cu-Kα radiation on 200 μm and 30 μm Ni-Mn-Ga foils (prepared by ED method).

Fig. 5.1 reveals that foils are exhibiting different crystal structure for the bulk Ni-Mn-Ga alloy (10M) and 30 μm (14M) foil. In addition, the interatomic spacing (d) changes form 1.486 Å to 1.378 Å while decreasing the foil thickness from 200 μm to 30 μm. These XRD results are correlated with the influence of surface damage on the crystal structure investigated for bulk Ni-Mn-Ga alloys [92]. The XRD results show that the foils prepared by electrical discharge machining show different crystal structure especially for the foils below 200 μm thickness. Further experiments are carried out to study the influence of surface damage on the phase transformation, twinning stress and ultimately magnetic field induced reorientation of Ni-Mn-Ga foils.

### *5.1.2 Thermal shape memory characteristics*

The phase transformations characteristics of Ni-Mn-Ga foils are investigated by using differential scanning calorimetry (DSC). Fig. 5.2 shows the DSC characteristics of a 80 μm thick single crystalline Ni-Mn-Ga foil in as-received condition. The DSC curve is characterized by analyzing an exothermic peak on cooling corresponding to the phase transformation of martensite ($M_p$ = 62 °C) and an endothermic peak during heating corresponding to the reverse phase transformation to austenite ($A_p$= 71 °C), respectively.
The thickness dependence phase transformation temperatures of Ni-Mn-Ga foils are listed in Table 5-1 as shown below. The phase transformation temperatures of Ni-Mn-Ga foils ranging from 200 - 30 μm reveal a linear shift in austenite phase temperature from 62 °C for 200 μm to 84 °C for 30 μm and also for martensite phase temperature from 54 °C for 200 μm to 74 °C for 30 μm, respectively. The $T_c$ value does not show significant change and remains constant while decreasing the foil thickness. Hence, DSC results confirm the significant increase of martensite transformation temperature while decreasing foil thickness. However, this cannot be explained by a small variation of chemical composition. Instead, these results indicate that the foil preparation methods introduce a considerable intrinsic stress that can cause significant increase in martensite temperature upon decreasing the foil thickness. It might

be possible due to the significant amount of defects (lattice defects such as stacking faults, dislocations and interstitials) introduced into the foil surface that may cause an increase of martensite temperature because of stress assisted phase transformation.

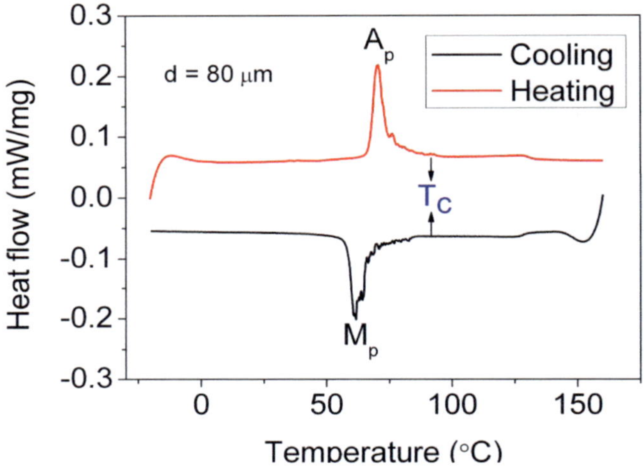


**Figure 5.2:** Phase transformation characteristics of a 80 μm thick single crystalline Ni-Mn-Ga foil investigated by DSC.

| Thickness (μm) | Austenite ($A_p$) (°C) | Martensite ($M_p$) (°C) | $T_c$ (°C) |
|---|---|---|---|
| **30** | 84 | 74 | 95 |
| **60** | 82 | 73 | 95 |
| **80** | 71 | 62 | 96 |
| **200** | 62 | 54 | 97 |

**Table 5-1:** Phase transformation temperatures of Ni-Mn-Ga foils in the thickness range of 30 - 200 μm as measured by DSC.

In this case, surface damage of the 30 μm thin foil is higher than of the 200 μm thick foil. In bulk Ni-Mn-Ga alloys, surface damage due to the alloy preparation method does not reveal a shift of phase transformation temperatures because the area of surface damage attributed with respect to the volume of the Ni-Mn-Ga alloy is less [92]. The type and role of the defects contributing to the temperature shift would need further quantitative microstructural investigation.

### *5.1.3 Mechanical properties*

The mechanical performance of as-received Ni-Mn-Ga foils is investigated by tensile tests performed in strain-controlled mode. Fig. 5.3 shows stress-strain characteristics determined by tensile loading on as-received single crystalline foils. The investigations are carried out on foils having 200, 60 and 30 μm thicknesses. An initial elastic response occurs below 0.5 % strain. Upon further loading, quasi-plastic deformation is observed as shown in Fig. 5.3.

The twinning stress is determined by using the tangential method. The twinning stress $\sigma_{tw}$ for the 200 μm thick foil is determined to be 5.5 MPa. It is 2.5 times higher than in the original bulk Ni-Mn-Ga alloy exhibiting 1.1 MPa at RT in compressive mode [89]. A quasi-plastic strain of about 4 % due to variant reorientation is observed and at this point, loading is stopped to avoid damage to the foil. The maximum strain of about 6 % and twinning stress of 5.5 MPa indicates that the foil of 200 μm thick is exhibiting 10M modulated martensite and it is similar to the bulk reference sample. In contrast to this, the 60 μm foil is exhibiting a high twinning stress of 17 MPa and a large plastic strain of about 10 %, which cannot be due to 10M modulated structure as of bulk reference sample. Instead, large strains in the order of 10 % are expected to occur for 14M or non-modulated martensite microstructures. The sudden stress drops as observed in the characteristic of 60 μm foil can be interpreted as serrations, since deformation in the surface layers are mostly accompanied by twin boundary pinning sites.

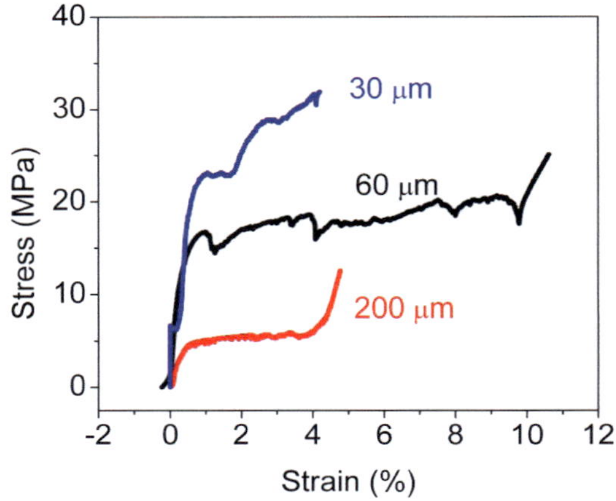


**Figure 5.3:** Stress-strain characteristics of as-received single crystalline foils measured at room temperature (RT).

In addition, stress-strain characteristics of the 30 μm foil reveals a high twinning stress of 23 MPa along with step type increase of stress as shown in Fig. 5.3. The experiment is continued up to 4 % plastic strain. Further loading is prevented in order to keep the foil for further measurements. From Fig. 5.3, it is obvious that the twinning stress is increasing as a result of the foil preparation process. Obviously, mechanical polishing introduces defects that further impede the mobility of twin boundaries. The high values of twinning stress observed for low foil thickness reveal that in as-received condition Ni-Mn-Ga foil actuation is not possible in a magnetic field [95].

The twinning stress is plotted versus the inverse thickness (1/d) for a series of tensile tests in the range of 30-200 μm Ni-Mn-Ga foils. Fig. 5.4 shows a continuous increase of twinning stress while decreasing the thickness of the foil from bulk Ni-Mn-Ga (5 mm) to 30 μm (least thickness).

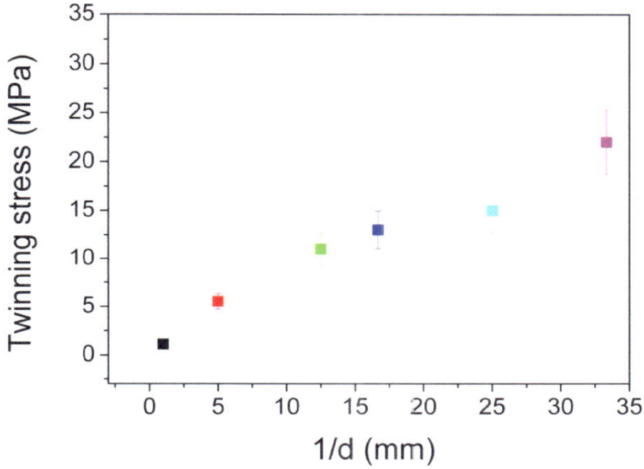


**Figure 5.4:** Twinning stress versus 1/d (d=thickness) feature of Ni-Mn-Ga alloy measured by tensile experiments.

The increase of twinning stress for decreasing the foil thickness is studied by the influence of defect surface layer on the twin boundary mobility. Since decreasing the foil thickness is a sequential process step involving mechanical polishing, the degree of damage attributed to the thin foil is large. This can be seen in Fig. 5.5, where the surface of the 200 µm foil shows almost homogeneous surface revealing single variant state. For 30 µm foil, the surface shows quite a complex morphology, where the twin pattern is localized randomly.

The surface damage in Ni-Mn-Ga foils due to mechanical polishing is illustrated by the schematic shown in Fig. 5.6. The deformed top layer on the foil surface indicates mechanically influenced layers with defects; below is a defect-free region not affected by surface treatments. In addition, twins with preferred orientation are indicated as twin boundary resting positions during tensile deformation experiment. These twin boundaries can move until they get stopped at defects in the surface layer. Twin boundaries can move through the foil only in small steps before they get twinned at defects in the surface layer or at local stress concentrations.

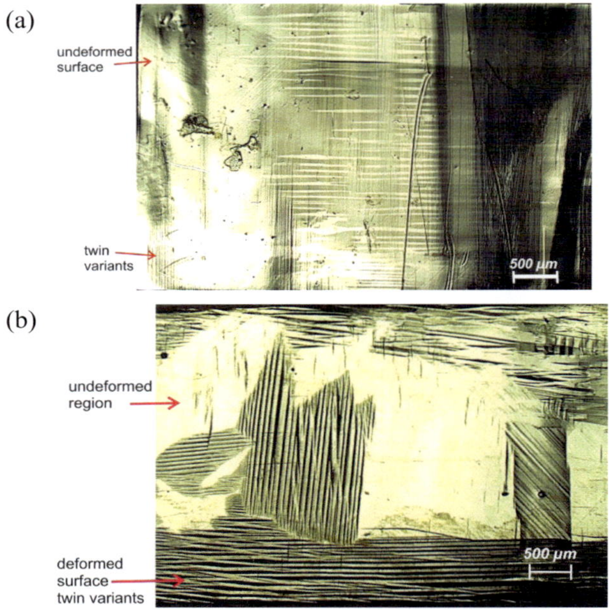


**Figure 5.5:** Optical micrographs showing the surface of (a) 200 µm Ni-Mn-Ga foil and (b) of a 30 µm Ni-Mn-Ga foil. The surface features are indexed.

The defects and local stress concentrations act as finely dispersed pinning sites at which twin boundaries get stopped. Thus, each pinning event leads to increase of stress and ultimately twinning stress increases [92, 96]. Hence, the foils prepared by ED method significantly influence the foil microstructure and hence twinning stress is influenced depending on the degree of damage.

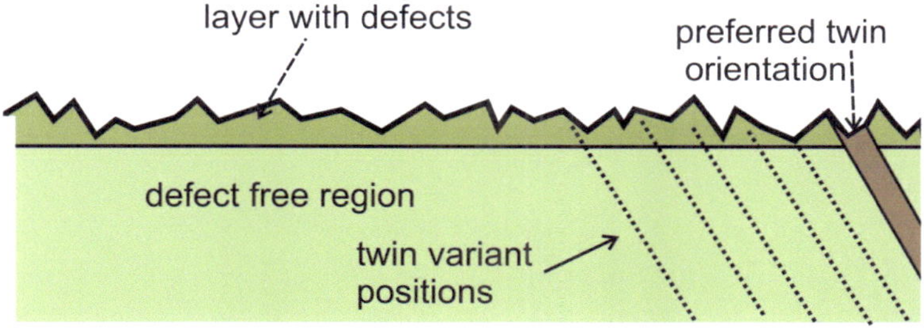


**Figure 5.6:** Schematic illustrating the effect of a deformed surface layer on the twin microstructure.

The drastic increase of twinning stress with decreasing the foil thickness is in agreement with change of phase transformation temperatures and crystal structure. However, these qualitative investigations further needs to be confirmed by probing the microstructure of the defected surface layers.

### *5.1.4 Microstructure of Ni-Mn-Ga foils*

To investigate the microstructure of the damaged surface layer, experiments are carried out to study the cross-section of the surface damage layer by using Focused Ion Beam (FIB). The Ni-Mn-Ga foil of 30 μm thickness is cut along the cross-section by using FIB. The depth of the FIB cut is 20 μm and cross-section is imaged with secondary electrons. Fig. 5.7 shows the cross-section of the damaged surface layer microstructure. It can be observed that the foil is exhibiting the homogeneous depth profile without any evidence of defect microstructure or deformed surface microstructure. Microscopically, FIB results confirm the defect-free microstructure along the cross-section of the foil. Further, experiments are carried out to explore possible defect contributions at the nanoscale by using transmission electron microscopy (TEM).

**Figure 5.7:** Cross-sectional SEM image of a Ni-Mn-Ga foil of 30 μm thickness. The surface and cross-section are indicated.

Fig.5.8 (a) shows a HR-TEM image of a Ni-Mn-Ga thin lamella prepared from the surface cross-section for the foil of 30 μm thickness. A distinct periodic twin variant structure along the both sides of the twin boundary is observed as indicated by red and green colors. A selected area diffraction pattern along both sides of the twin boundary indicate that twin variant plates are superimposed onto each other with an average width of the twin variant plate is 30 nm. Besides the periodic twin structure, an anomaly of different large twin variant plates is observed, which could break the symmetry of the twin boundaries. This might be due to deformed surface layer damage attributed by the foil preparation method. Electron diffraction pattern along [110] zone axis of Ni-Mn-Ga unit cell probed on these meso-scopic twins reveals no modulated crystal structure. The experiments are performed on different regions of the TEM lamella, but none of them shows a modulated 10M structure as expected from bulk reference samples [89].

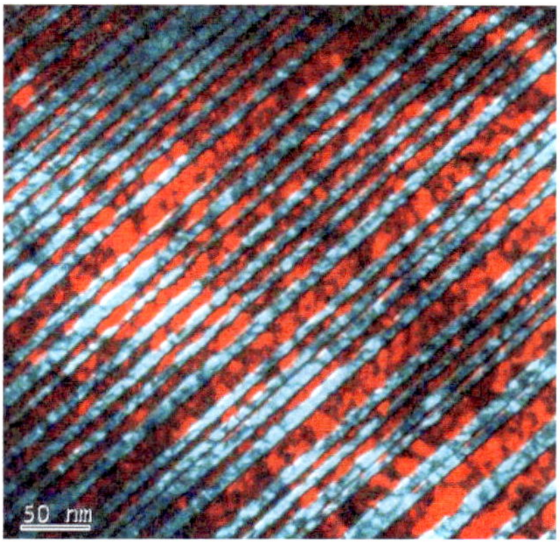


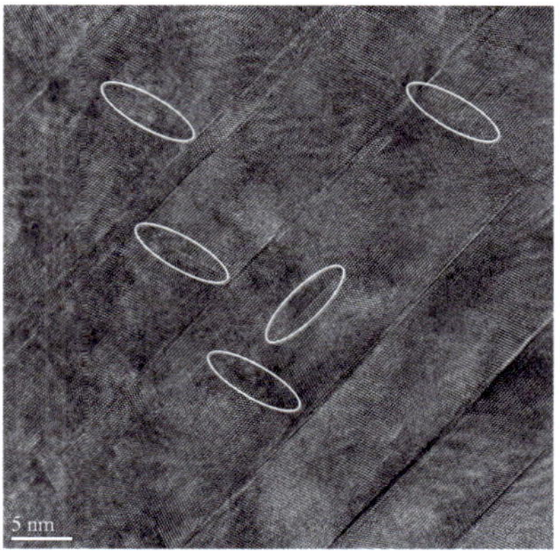


**Figure 5.8:** (a) STEM image of a Ni-Mn-Ga thin lamella along the cross-section of the foil indicating the mesoscopic twin variants highlighted by two different colors, (b) TEM image with high magnification showing stacking faults as indicated.

Fast fourier transforms (FFTs) mapped along both sides of the twin boundary indicate that twin variants are inclined by 45 degree with respect to one other. TEM results confirm the FCC cubic crystal structure with lattice constant a = b = c = 3.67 Å. The interatomic spacing between atomic planes is calculated along [111] direction to be $d_{111}$ = 2.12 Å. Hence these results confirm that the surface layer is showing an entirely different crystal structure instead of a modulated 14M or 10M martensite structure.

Fig. 5.8(b) shows a high magnification image of martensite twin variants revealing the defect structure constituting of both lattice dislocations and stacking faults. The disconnections across the interface are ascribed to a long-range strain field and shape change induced by the foil preparation method. In addition, stacking faults change the sequence of atomic layers as shown in Fig.5.8(b). Thus twin boundary mobility needs higher energy because the defects inhibit variant reorientation. This is in agreement with the higher twinning stress associated with the thinner foils compared to the thick foils. These lattice defects can act as pinning sites for the twin-boundaries.

The stress-strain measurements (Fig. 5.3) reveal that the impact of surface damage on the twin mobility is higher for thin foil. This is because of the large amount of defect structure contributing to the twin boundary mobility. Hence for the 30 μm Ni-Mn-Ga foil, large number of crystal defects (stacking faults and point defects) implies large surface damage caused by foil preparation. Moreover, as shown in Fig. 5.8(b) each single twin domain in the nano-scale is influenced by the lattice defects. This is different from the situation of single domain crystal with defect-free, electro polished surfaces, where the twinning stress is nucleation controlled at the onset of deformation [92]. Plastic deformation of the surface layer that creates pinning sites for twin boundaries results not only in increasing the twinning stress but also change the maximum strain as distinguished from stress-strain measurements (Fig. 5.3). In summary, the surface damage affecting the crystal structure and microstructure for 200 and 30 μm Ni-Mn-Ga foils is confirmed by HR-TEM. In addition, phase transformation characteristics along with mechanical properties supports increasing influence of surface damage for decreasing foil thickness.

Modification of the near-surface microstructure can be tailored to achieve the desired twinning stress and to observe the magnetically induced actuation. To achieve low twinning stress and observe magnetic field induced reorientation in Ni-Mn-Ga foils, further experiments have been carried out in Ni-Mn-Ga foils to evaluate and minimize the surface damage. In bulk Ni-Mn-Ga alloys, surface damage was extensively investigated by grinding, spark erosion, laser cutting, cold forging, ion implantation and abrasive wearing methods by Chmielus et al [92, 94]. They showed that surface damage layer can be removed or minimized by electro polishing and by repeated mechanical training in rotating magnetic field.

### *5.1.5 Training of Ni-Mn-Ga foils*

By thermo-mechanical training (chapter-3) of single crystalline foils of 200 and 80 μm thickness, the twinning stress is significantly reduced. Fig. 5.9 shows a stress-strain characteristics measured in tensile mode along the length direction of the foil specimen having 200 μm thickness. Before this measurement, a magnetic field of 0.85 T is applied perpendicular to the length direction of the Ni-Mn-Ga foil to induce c-axis orientation of the martensitic variants along the applied magnetic field. A very low twinning stress of about 0.3 MPa is observed, which is comparable to the results of the corresponding Ni-Mn-Ga bulk single crystal [89]. Therefore, in conclusion, the tensile training method allows the complete recovery of the low twinning stress of the bulk reference sample.

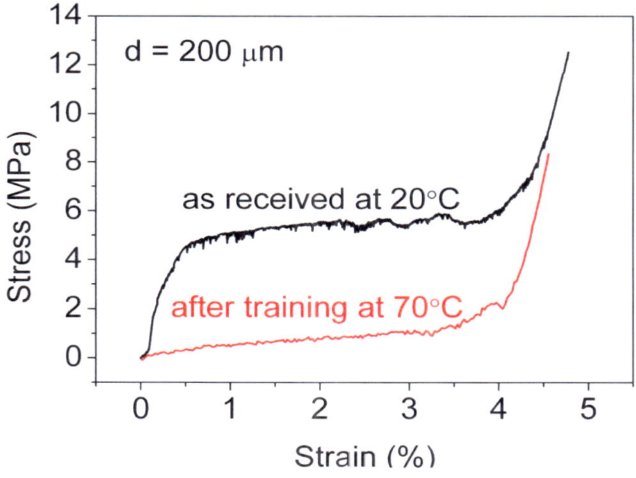


**Figure 5.9:** Stress-strain characteristics of 200 μm thick Ni-Mn-Ga foil before and after thermo-mechanical training.

Similarly, the thermo-mechanical training is performed on the 80 µm foils. Two different single crystalline foils are prepared for uniaxial tensile loading. One is as-received single crystal and the other is a thermo-mechanically trained one. Uniaxial multi-step tensile loading is conducted for both Ni-Mn-Ga foils at room temperature as shown in Fig. 5.10.

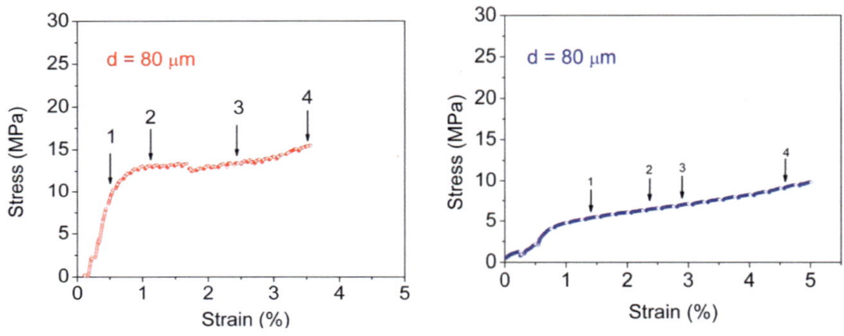


(a) (b)

**Figure 5.10:** Stress - strain characteristics of (a) as-received and (b) thermo-mechanical trained single crystal Ni-Mn-Ga foil under uniaxial tensile loading. 1, 2, 3, 4 numbers indicate strain points where DIC is measured.

It is apparent from Figs. 5.10 (a) and (b) that, there is a large difference in twinning stress after thermo-mechanical training. The twinning stress is estimated to be 13 MPa for as-received specimen and 5 MPa for trained specimen respectively. Initial stress value upto 0.5 % strain in Fig. 5.10 (b) is attributed to elastic behavior of sample fixation parts. Fig. 5.10 shows that thermo-mechanical training allows to reduce the twinning stress, but it is still higher than the bulk reference value. This might be due to the higher degree of surface damage in the 80 µm foil as compared to the 200 µm foil. This is in agreement with the change of the crystal structure observed in the XRD measurements. Moreover, it is also confirmed by a change in the phase transformation characteristics as described before.

The strain field distribution arising in two single crystal Ni-Mn-Ga foil specimens is measured using the DIC method [87, 97]. Fig. 5.11 shows the local strain distribution of as-received and trainied Ni-Mn-Ga foil of 80 μm thickness under multi-step tensile loading. The strain field distribution is determined within a small region on the Ni-Mn-Ga foil. Each strain distribution is calculated for a tensile displacement of 0.02 mm. For the as-received Ni-Mn-Ga foil, the deformation behavior is inhomogeneous with intermittent strain bands that are appearing and disappearing during the course of deformation. In contrast, the trained Ni-Mn-Ga foil shows a more homogeneous deformation behavior all over the Ni-Mn-Ga foil surface.

XRD investigation performed on the 80 μm thick Ni-Mn-Ga foil shows a 14M martensitic crystal structure at room temperature. In the present uniaxial deformation experiments (Fig. 5.10), the twinning stress of as-received Ni-Mn-Ga foil is much larger than that of trained Ni-Mn-Ga foil.

It is evident that after thermo-mechanical training of Ni-Mn-Ga foils, redistribution of martensite variants occurs by preferential alignment of the long a-axis along the tensile loading direction. Single crystals having a high fraction of defect states are more likely to show the macroscopic and microscopic inhomogeneous deformation. In fact, the strain distribution results show local deformation bands that appear during deformation of as-received Ni-Mn-Ga foil. It is assumed that the defects rearrange due to stress assisted thermo-mechanical training resulting in a preferential orientation of the variants all over the Ni-Mn-Ga foil specimen.

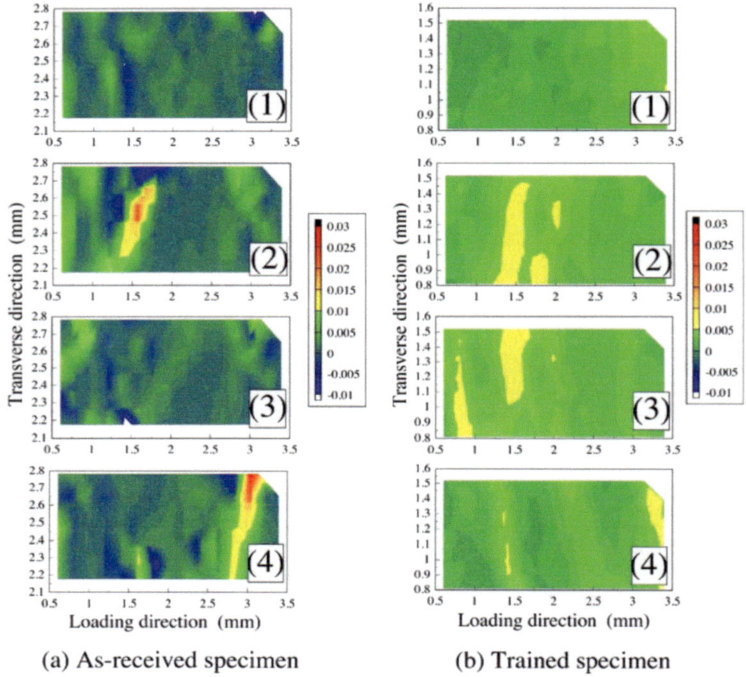


**Figure 5.11:** Longitudinal strain distribution of as-received and trained single crystal Ni-Mn-Ga foil under uniaxial tensile loading. The images 1,2,3 and 4 are calculated for each strain point shown in Fig. 5.10.

### *5.1.6 Magnetic field induced reorientation*

The magnetic field induced reorientation in Ni-Mn-Ga foil actuators is investigated in a homogenous magnetic field by directly observing the deflection of the freely movable end of the actuator by a video microscope. The magnetic field is ramped step-wise between 0 and 0.85 T, while applying the field direction either parallel ($H_x$) or perpendicular ($H_y$) to the long axis of the foil actuator as shown in Fig. 5.12(a). Below the critical magnetic field, the magneto strain keeps below the resolution of the measurement set-up of about 3 μm. Upon further increase of the magnetic field, the actuator suddenly contracts by about (Δx) 35 μm and than remains deformed until the maximum field of 0.85 T is reached. No reversible effect is observed when the magnetic field is switched off.

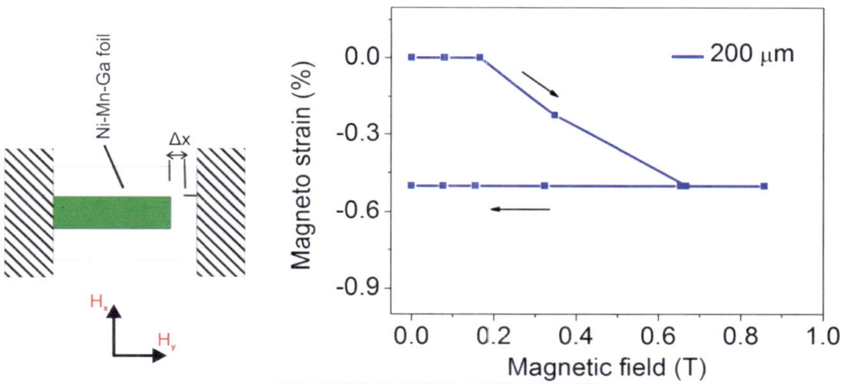


(a) (b)

**Figure 5.12** (a) Schematic of the magneto-strain measurements, (b) Magneto-strain characteristics of single crystalline Ni-Mn-Ga foil having 200 μm thickness.

The corresponding magneto-strain characteristics are shown in figure 5.12 (b). The contraction of 35 μm corresponds to about 0.5 % magneto strain. Thus, about 10 % of the variants contribute to the magneto-strain effect. Partial reorientation is observed in magneto-strain experiments in comparision with 6 % strain in bulk Ni-Mn-Ga alloy. One reason for the partial reorientation effect is that tensile training only acts on one of the two long a-axis and thus only allows partial prealignment of the martensite variants. Consequently, only half of all martensite variants can reorient. This kind of prealignment favors a unique, parallel alignment of the twin boundaries. Hence, intersections of the twin boundaries may act as pinning centers, hindering the twin boundary mobility. Furthermore, pinning of twin boundaries at internal defects cannot be excluded.

Depending on the deformation energy required to overcome the pinning of the twin boundaries, reorientation may not be finished. Additional application of a biasing stress may help to increase the fraction of reorienting martensite variants, which remains to be tested in further experiments. An even better prealignement of martenstie variants is expected to be achieved by thermo-magneto-mechanical training. In particular, an additional magnetic field applied perpendicular to the tensile loading direction is expected to favor an initial single variant state.

In conclusion, single crystalline Ni-Mn-Ga foil actuators show an initial large twinning stress and thus no measurable magneto strain due to impact of thickness reduction. It is shown, that the low twinning stress of the bulk reference actuators is recovered after thermo-mechanical training in tensile mode. However it has a clear limitation due to large fraction of defect density in thinner foil, yielding non-recoverable low twinning stress by training. Hence further improvement of the foil microstructure is desired as well as novel training methods will be required to achieve further improvement of performance.

## 5.2 Actuation of Ni-Mn-Ga foils

Here, Ni-Mn-Ga foil actuators are developed by the use of ferromagnetic shape memory effect. Two Ni-Mn-Ga foil stripes having 200 and 30 μm thickness are investigated. Single crystalline foil of 30 μm thickness is the least possible thickness so far fabricated by the abrasive wire saw-cutting method. In contrast to the foils prepared by electric discharge machining, this new technique allows to prepare Ni-Mn-Ga foils with considerably less damage to the surface layers.

### *5.2.1 Material properties*

The crystallographic structure of the Ni-Mn-Ga foils is investigated by X-ray diffraction (Cu-Kα radiation, wavelength λ = 1.54 Å). Fig. 5.13 shows an x-ray diffraction measurement after foil preparation and crystal orientation by a magnetic field.

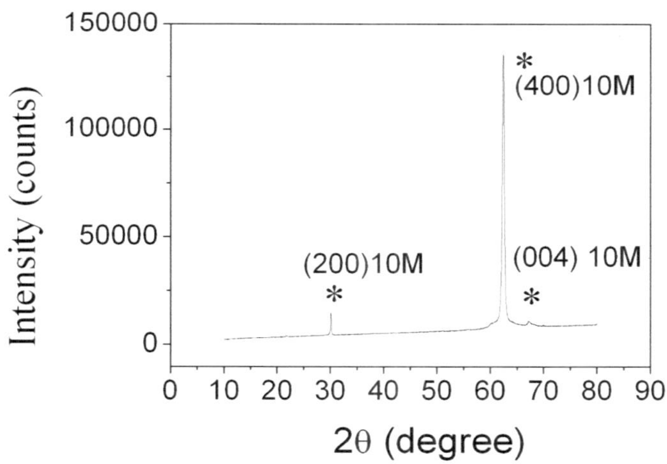


**Figure 5.13:** X-ray diffraction measurement of a single crystalline Ni-Mn-Ga foil of 200 μm thickness.

One major sharp peak is observed at 62.419° caused by (400) reflection indicating that the foil is almost in a single variant state. The peak is due to the (400) reflection of the 10M martensite structure at room temperature. The structure of the 10M martensite has been determined to be tetragonal [98, 99]. In multi-variant state, the material consists of layered mesoscopic regions of twin variants separated by twin boundaries [100].

The phase transformation characteristics of 30 μm thick Ni-Mn-Ga foil are investigated by temperature dependent electrical resistance measurement. The measurement reveals the typical hysteresis of phase transformation between austenite and martensites [99, 101] as shown in Fig. 5.14. Upon cooling from high temperature, a sharp increase of resistance is observed at 54 °C indicated as martensite start temperature. The martensitic finish temperature is determined to be 50 °C.

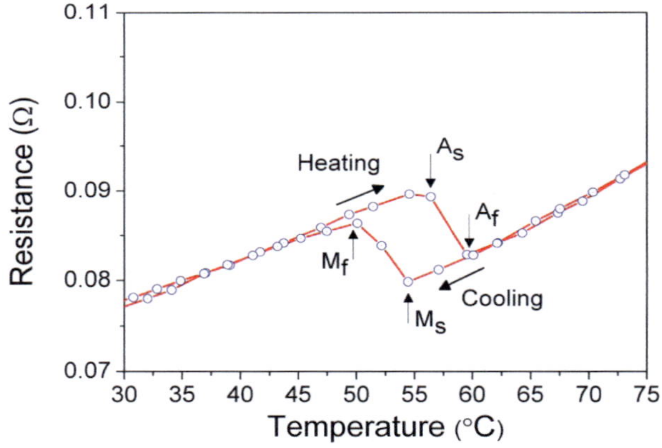


**Figure 5.14:** Temperature-dependent electrical resistance characteristics of Ni-Mn-Ga foil of 30 μm thickness.

Similarly, upon heating a sharp decrease of resistance is observed at 55 °C indicating the austenite start temperature. The austenite finish temperature is determined to be 59 °C. The measured phase transformation temperatures for 30 μm Ni-Mn-Ga foil deviate by about 2 °C from the corresponding bulk Ni-Mn-Ga phase transformation temperatures. This small difference of 2 °C is within measurement accuracy. Hence, these results confirm that the Ni-Mn-Ga foils are almost not affected by the foil preparation method.

### *5.2.2 Mechanical properties in a magnetic field*

An engineering stress-strain characteristic of the Ni-Mn-Ga foil is shown in Fig. 5.15(a). The experiment is performed by a tensile testing machine in strain-control mode. Before the measurement, a magnetic field of 1 T is applied along the tensile direction ($H_x$) to adjust to a single variant state with c-axis orientation being parallel to the tensile direction.

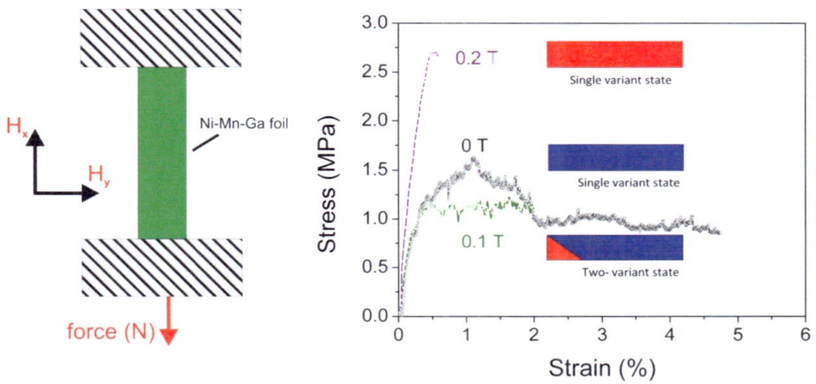


**Figure 5.15:** (a) Schematic of the tensile experiment in counteracting magnetic fields, (b) tensile stress-strain characteristics of a single crystalline Ni-Mn-Ga foil of 200 µm thickness in single-variant and two-variant states as indicated.

Alternatively, an additional field of 0.1 T is applied in transverse direction ($H_y$) after adjustment of the single variant state to set up a two-variant state with a small fraction of variants having a c-axis orientation in transverse direction. A magnetic field is applied to the Ni-Mn-Ga foil along both $H_x$ and $H_y$ direction, $H_x$ is the magnetic field applied along the length direction of the sample and $H_y$ is along the width direction of the sample as shown in Fig. 5.15(a).

An initial linear elastic response is observed up to a strain $\varepsilon_{tw}$ of about 0.3%. The corresponding Young's modulus $E_{tw}$ is estimated to be about 400 MPa. In single variant state, the onset stress needed to start twin boundary motion is about 1.6 MPa, while the stress needed for further reorientation of the variants is only about 1.2 MPa. In two-variant state, the onset stress disappears as only reorientation of the martensite variants occurs. This result is in-line with previous stress-strain experiments on bulk Ni-Mn-Ga samples loaded by compression [101]. After about 5 % relative elongation, the twin variants are reoriented and the stress increases subsequently (not shown here). The maximum strain is smaller than the maximum tetragonal distorsion $\varepsilon_{max}$. This is due to the constraints induced by fixation of both ends of the foil, which create mechanically inactive MSM regions and thus reduce the overall strain.

### *5.2.3 Actuation in counteracting magnetic fields*

Reorientation of martensite variants is achieved, e.g., by applying two counter-acting magnetic fields being orientated in-plane ($H_x$) and transverse direction ($H_y$) of the Ni-Mn-Ga foil stripe. Fig. 5.16 shows a schematic of a beam shaped Ni-Mn-Ga foil, which is mounted at one end, while the other end is freely movable. The microstructure consists of almost single variant of tetragonal martensite having a short c- and two long a-axes as sketched in the insets. The short c-axis coincides with the magnetic easy axis. Therefore, the martensite variants with c-axis orientation along the direction of a magnetic field are energetically favored. Thus, by applying a magnetic field in transverse direction $H_y$ (Fig. 5.16(a)), the short c-axis aligns in y-direction and vice versa for $H_x$. The maximum possible shape change is given by completely reorienting the crystal from a single variant state with initial c-axis orientation being in y-direction to a single variant state with c-axis orientation being in x-direction.

Beam-shaped Ni-Mn-Ga foil actuators are fabricated by chemical micromachining and subsequent bonding onto an alumina substrate [12]. An opening is provided into the substrate to enable frictionless movement. Before the measurements, the martensite variants are prealigned by a magnetic field to be in a single variant-state with c-axis orientation along the transverse (y-) direction. While increasing the magnetic field along the tensile direction, the deflection of the movable end of the foil actuator is monitored by a video microscope. Relative movements of the setup are eliminated by using a reference marker close to the freely movable end of the actuator, which is fixed to the same substrate. No external mechanical force is applied in these measurements.

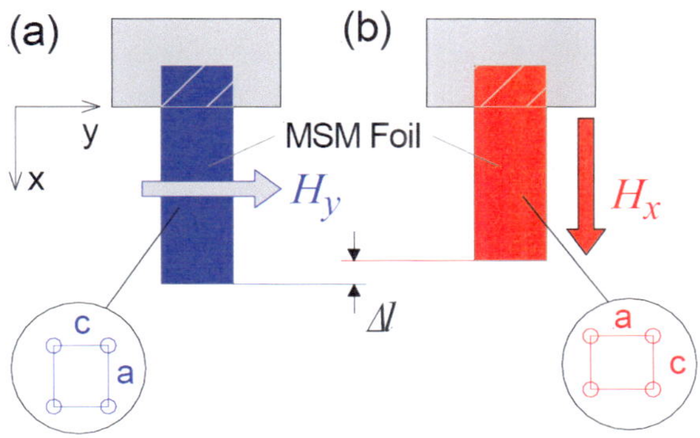


**Figure 5.16:** Principle of MSM actuation by two counter-acting magnetic fields $H_x$ and $H_y$ in a single crystalline Ni-Mn-Ga foil. The crystallographic orientations are indicated in the insets with a and c denoting the long a- and short c-axis, respectively.

Fig. 5.17 shows typical magneto-strain actuation characteristics measured on Ni-Mn-Ga foils having 200 μm and 30 μm thickness. At small magnetic fields below 100 mT, the magneto-strain stays almost constant. The critical field $H_c$ to induce the MSM effect for 200 and 30 μm is calculated by using the tangential method to be 120 and 190 mT respectively, which is in-line with previous magnetization experiments on similar foil specimens [16]. The high critical field of 190 mT exhibited by the 30 μm foil as compared to the 200 μm Ni-Mn-Ga foil is contrary to the prediction of decrease in critical field with decreasing the size of Ni-Mn-Ga actuator [16]. In order to evaluate the size dependence of the critical field, further experiments are necessary on Ni-Mn-Ga foil actuators having various thicknesses. The increase in critical field for the 30μm foil could be attributed to the contribution of defects in the surface layers created by foil preparation. The actuator movement continues until the final contraction is reached at about 400 mT for both foils respectively. The final strain achieved on both Ni-Mn-Ga foils is estimated to be ~5.9 %, which is somewhat smaller than the theoretical limit of 6 % strain. This shows that complete reorientation occurs in most parts of the Ni-Mn-Ga foil, which is now in a single-variant state with c-axis orientation along the x-direction.

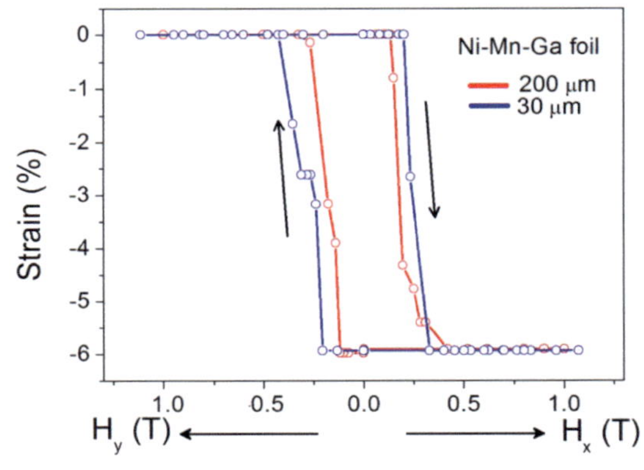


**Figure 5.17:** Magneto-strain characteristics of a beam-shaped Ni-Mn-Ga foil actuator of 200 and 30 μm thickness at room temperature.

Due to the small thickness of the Ni-Mn-Ga foil actuator compared to the lateral dimensions, the effect of demagnetization can be neglected when considering the in-plane magneto-strain performance. No further movement is observed when increasing the field further up to 1.5 T. The reorientation is irreversible and remains constant after decreasing the magnetic field to zero, because no external force is applied. By applying the magnetic field subsequently in transverse direction (y-direction), the initial single-variant state is restored. In conclusion, Ni-Mn-Ga foils are suitable candidates for linear actuation application in counteracting magnetic fields with large strain up to 6 % by variant reorientation.

### *5.2.4 Linear actuation of Ni-Mn-Ga foils*

For applications, the load-dependent performance of Ni-Mn-Ga foil actuators is of special interest. Hence in order to develop a practical linear actuator, it is convenient to replace the transverse magnetic field by a mechanical load. Single crystalline Ni-Mn-Ga bulk actuators usually contain a compressive loading setup [102, 103]. For a Ni-Mn-Ga foil actuator, however, this kind of loading is not possible as compression would cause buckling. Instead, a tensile load along the long side of the specimen is applied as sketched in Fig. 5.18. In this case, one of the long a-axis (10M) favorable orients in tensile direction (Fig. 5.18 (a)). By applying a magnetic field along the tensile direction, again a single-variant state with final c-axis orientation being parallel to the magnetic field is obtained (Fig. 5.18 (b)). The tensile load-dependent magneto-strain characteristics determined for 200 and 30 μm Ni-Mn-Ga foils are summarized in Fig. 5.20. The loading setup is sketched in Fig. 5.19. It mainly consists of a fixation unit at the end of the Ni-Mn-Ga foil, a guiding wire and a miniature weight to guarantee for displacement-independent loading.

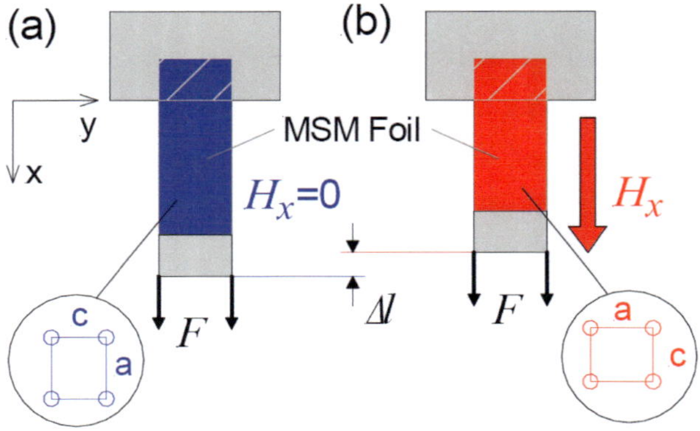


**Figure 5.18:** Principle of MSM actuation in the presence of a tensile load and a magnetic field $H_x$ in a single crystalline Ni-Mn-Ga foil. The crystallographic orientation are indicated.

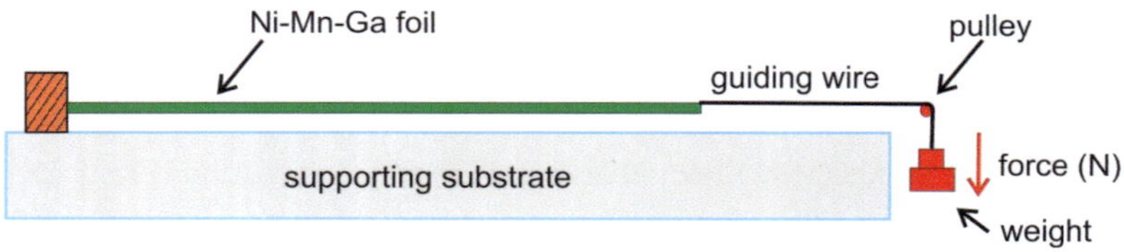


**Figure 5.19:** Schematic of the loading setup for linear actuation of Ni-Mn-Ga foils.

By increasing the magnetic field $H_x$, the movable end of the Ni-Mn-Ga foil contracts, while the weight is lifted. Thereby, tensile load guides the direction of linear actuation. In addition, it acts as a reset element to attain the initial position once the magnetic field is switched off. In this configuration, the back and front end of the Ni-Mn-Ga foil are constraint due to fixation and loading, respectively. Therefore, measurement of the total length change gives rise to smaller overall strain denoted as the effective strain $\varepsilon_{eff}$, which results from both mechanically inactive and active MSM regions. At zero load, the resulting effective strain is about ~ 4.4 % for both 200 and 30 μm foils, see Fig. 20.

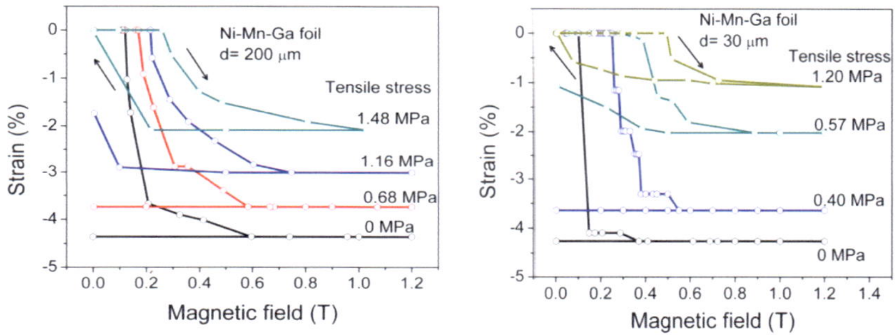


**(a)** **(b)**

**Figure 5.20:** Tensile load dependent magneto-strain characteristics of a beam shaped Ni-Mn-Ga foil at room temperature (a) Ni-Mn-Ga foil having 200 μm (b) Ni-Mn-Ga foil having 30 μm thickness.

For the 200 μm foil and tensile stress range below 1.48 MPa, resetting of linear actuation in Ni-Mn-Ga foil remains incomplete, i.e., a stress-dependent residual strain remains after the magnetic field is switched off. In this case, an additional transverse magnetic field is applied to fully recover the initial state of c-axis orientation being in y-direction. For increasing tensile stress, the critical magnetic field increases as the magnetic field energy has to overcome the additional stress barrier. In addition, the magneto strain decreases, which is in line with results obtained for bulk Ni-Mn-Ga single crystals loaded by compression [23, 24]. At 1.48 MPa, tensile loading is sufficient to reset the linear actuation in Ni-Mn-Ga foil having 200 μm thickness as shown in Fig. 5.20(a). In this case, repeatable full actuation cycles are performed with a maximum actuation stroke of 2.2 %, no transverse magnetic field is required to reset the linear actuator. However, tensile loading does not necessarily result in the initial single variant state of c-axis orientation along y-direction (Fig. 5.18 (a)), which further limits the achievable net displacement. Up to now, the effects of different constraints given by fixation and reset mechanism have not been studied systematically. This will be important for the future optimization of layout and the development of design rules.

Tensile load-dependent magneto strain characteristics is also investigated for 30 μm Ni-Mn-Ga foil as shown in Fig. 5.20(b). From the magneto-strain characteristics, decrease in magneto-strain is observed with increase in tensile load and the switching field increases with increase in counteracting tensile load. This behavior is similar to the characteristic features observed for the 200 μm Ni-Mn-Ga foil. In contrast to the 200 μm Ni-Mn-Ga foil, the ultimate magneto-strain for 30 μm foil depends very strongly on stress in the range between 0.40 and 0.57 MPa, see Fig. 5.20(b). The high critical field of 0.6 T is applied to reset the actuator to the original position. At 1.2 MPa, fully repeatable actuation cycles are performed at maximum actuation stroke of 1 % strain. This peculiar feature further needs to be understood in more detail by considering boundary conditions (fixation, loading) and size effect (foil thickness) on the performance characteristics of the Ni-Mn-Ga foil linear actuators. Despite the similar features on load-dependent linear actuation characteristics, these investigations give more insight concerning the optimum value of the critical stress and critical field for linear actuation of Ni-Mn-Ga foils in the thickness range of 30-200 μm.

In order to develop a portable linear actuator device as shown in Fig. 5.21, the magnetic field source needs to be integrated along with the foil actuator and reset element. Furthermore, a versatile reset mechanism is required. So far, the presented results on linear actuation of Ni-Mn-Ga foils have been obtained for an external solenoid.

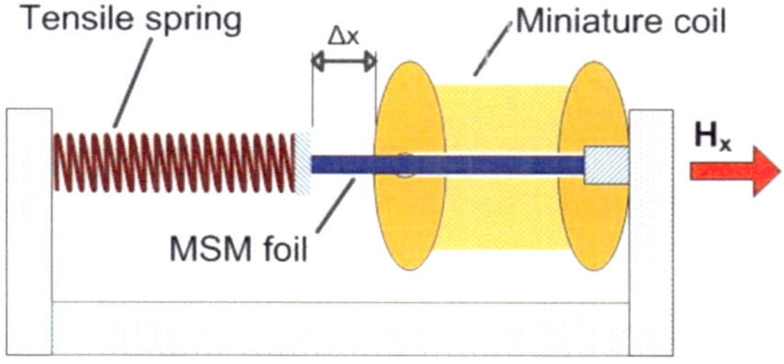


**Figure 5.21:** Layout of a linear actuator consisting of a Ni-Mn-Ga foil loaded by a tensile spring and a miniature coil. The displacement of the moving element between Ni-Mn-Ga foil and biasing spring is Δx.

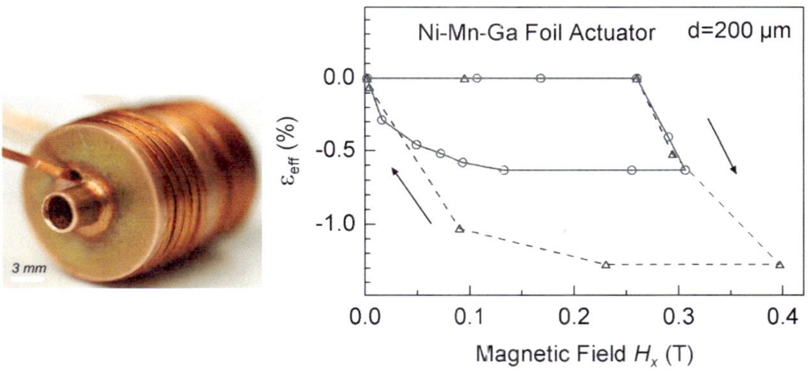


**Figure 5.22:** Demonstrator of a miniature coil (diameter of 19 mm) to operate the MSM actuator in pulsed mode (a); Magneto strain characteristics for maximum magnetic fields of 0.3 and 0.4 T (b).

The main components of this type of linear actuator are, besides the Ni-Mn-Ga foil, a miniature external coil to generate the in-plane magnetic field and a tensile reset spring. The force of the tensile spring is adjusted to generate a maximum tensile stress of 1.5 MPa. Fig. 5.22(a) shows a demonstrator of a miniature coil with an air core adjusted for the Ni-Mn-Ga foil. It consists of a core with inside diameter of 3 mm and a copper wire with number of turns of 20 x 25. By pulsed operation at 33 A, the coil is capable of generating a homogeneous magnetic field $H_x$ of about 0.35T. The pulse duration is set to about 100 ms to avoid overheating. As the maximum magnetic field of this type of linear actuator is limited, only partial reorientation of martensite variants can be achieved. Fig. 5.22(b) shows that the corresponding maximum magneto strain at 0.4 T is about 1.3 % for the 200 μm thick Ni-Mn-Ga foil.

In summary, a reversible linear actuation is observed in the beam-shaped Ni-Mn-Ga foil actuators of 200 and 30 μm thicknesses due to the MSM effect. Direct optical measurements reveal a maximum strain response of 5.9 % above a critical magnetic field of 0.12 T. In a practical linear actuator, the maximum usable strain is limited by the constraints of fixation and loading. In the present case, the MSM linear actuator shows repeatable full actuation cycles with a maximum actuation stroke of 2.2 % when operated at a tensile load of about 1.5 MPa. The maximum actuation stroke of 6 % could be achieved by replacing the constant tensile loading of Ni-Mn-Ga foil with tensile elastic spring (shown in Fig. 5.21); here the constraint of loading and fixation is neglected. Tensile elastic spring can induce precise control of force opposing the Ni-Mn-Ga foil in a magnetic field. Hence reversible actuation is possible by suitable selection of spring constant and optimum pre-stress of spring, which can overcomes the twinning stress of Ni-Mn-Ga foil in a magnetic field.

Design rules for optimum fixation of the Ni-Mn-Ga foil at both ends are developed. Inactive twin regions might occur due to fixation, which block twin boundary motion and may result in increase of switching field for the onset of twin boundary mobility. This can be solved by fixing the Ni-Mn-Ga foil at the face side of a cross-section to allow for easy twin boundary mobility. In addition, sometimes this kind of fixation results in bending of the Ni-Mn-Ga foil in out-of-plane direction. This could increase the switching field due to competing twin boundaries. Hence keeping the actuator geometry planar by using guiding elements is desired. Furthermore, magnetic field sources have to be designed in such a way that a uniform homogeneous field distribution is being subjected to Ni-Mn-Ga foil. For small scale applications generating high magnetic field strengths requires high current densities, which might results in high temperature operating environment due to continuous operation of direct currents. Therefore, applying current in pulsed mode is desired for optimal performance of the actuator.

Finally, design rules to increase the work output of a foil actuator coupling with external load have to be developed. As discussed above, the reversible actuation strain in Ni-Mn-Ga foils can be increased with bias spring mechanism with optimized spring constant. However, the spring constant and linear elongation of the spring should match to the temperature dependent stress-strain characteristics of the Ni-Mn-Ga foil specimen in order to achieve a maximum actuation stroke in a magnetic field. The parameters for optimum actuation stroke will be determined including prestrain, prestress, driving current and spring constant in a biasing spring actuation mechanism. In addition, material inhomogenities have to be taken into account by assuming different values of twinning stress in different regions of the Ni-Mn-Ga foil actuator. Due to large deviation in material inhomogenities the reversible actuation strain of the actuator decreases because of the amount of magnetic field required to complete the variant reorientation increases [104]. Thus it will limit the actuator performance in a homogeneous magnetic field.

Despite the large strains, the work output of Ni-Mn-Ga foils is limited due to relatively low blocking stress of 2-6 MPa above which control by magnetic field is no longer possible [56]. To date, attempts to enhance the work output limitation have been mainly focused on changing the concentrations of the off-stoichiometric compounds of the MSM alloys to increase the magnetocrystalline anisotropy energy [105]. However, using Ni-Mn-Ga foils one could perform a systematic study of inter-relationship between Ni-Mn-Ga foil thickness, blocking stress and reversible actuation strain in order to understand the ways to increase the work output in more detail.

# Chapter 6

# Freestanding epitaxial Ni-Mn-Ga films

The development of Ni-Mn-Ga microactuators showing large reorientation strain requires the fabrication of epitaxial or highly orientated Ni-Mn-Ga films, which are at least partially free-standing. Single crystalline Ni-Mn-Ga films are prepared by epitaxial growth on a MgO substrate. Since the magnetically induced reorientation of martensite variants is blocked by a rigid substrate constraint, the films have to be released from the substrate after deposition. A process flow is presented to successfully release Ni-Mn-Ga films grown on a MgO (100) substrate by using Chromium as sacrificial layer. In addition, microstructural investigations are shown before and after variant reorientation to understand the effect of stress-induced reorientation on the martensite microstructure. Temperature-dependent material and mechanical properties are investigated. Finally, thermo-mechanical training is performed on the freestanding Ni-Mn-Ga films in order to decrease the twinning stress.

Three different types of Ni-Mn-Ga films are used for the investigation, differing by composition, deposition temperature and thickness as listed in Table 6-1. Thin films are sputter deposited using a $Ni_{46}Mn_{32}Ga_{22}$ alloy target. The operating pressure is maintained at 0.008 mbar, the substrate to target distance is fixed at 90 mm. Under these similar process conditions, films are deposited at different temperatures and sputtering rates. It is well known that film thickness, composition and deposition temperature play an important role in formation of the surface twin pattern, because these process parameters control the biaxial residual stresses and thus the martensite structure [81]. Three different compositions corresponding to each deposition temperature are identified. In Ni-

Mn-Ga films, crystal structure of martensite is extremely sensitive to composition; thus, the three types of samples may differ significantly.

| Film name | Composition | Deposition temperature | Thickness |
|---|---|---|---|
| Sample-1 | $Ni_{50.5}Mn_{30.5}Ga_{19}$ | 400°C | 1 µm |
| Sample-2 | $Ni_{47.7}Mn_{31.1}Ga_{21.2}$ | 300°C | 2 µm |
| Sample-3 | $Ni_{49.2}Mn_{32.4}Ga_{18.4}$ | 350°C | 2 µm |

**Table 6-1:** Ni-Mn-Ga films sputtered deposited on the MgO substrate with Cr (100nm) as a sacrificial layer at different temperatures.

## 6.1 Sacrificial layer technology

In recent years, considerable progress has been made towards the development of MSM microactuators. Dong et al [68, 74, 76] used molecular beam epitaxy (MBE) to deposit the epitaxial Ni-Mn-Ga films on a sacrificial $Sc_{0.3}Er_{0.7}As$ layer on GaAs (100) substrate, where it allowed preparing freestanding bridges and cantilevers. Up to now no other group repeated this approach, indicating that using MBE involves a complex process (e.g., low sputtering rate, difficult to control process) to deposit epitaxial Ni-Mn-Ga films. Sputtering has proven to be a reliable technique in order to obtain reproducible epitaxial films on various substrates. Khelfaoui et al [79] reported peeling of the sputtered epitaxial Ni-Mn-Ga film from a MgO substrate, but this technique is unreliable because of uncontrollable plastic deformation introducing dislocations and other defects into the film. Furthermore, the epitaxial growth of Ni-Mn-Ga is also possible on NaCl (100) substrates [14], which can be dissolved in water after deposition [106]. Though this approach appears to be quite simple, a NaCl substrate has several disadvantages. First, NaCl is hygroscopic and hence requires a careful bake-out to evaporate all water before deposition. Even small amounts of water result in the formation of a crater-like film morphology instead of the required continuous film. Second, adhesion of the metallic film to the NaCl substrate is poor. In particular for thick or stressed films, this sometimes results in delamination during deposition. Third, different epitaxial relationships have been observed: (001) and (110) orientation of Ni-Mn-Ga unit cell [106]. This indicates that reproducible epitaxial growth on this substrate may be difficult.

A novel technological approach of preparing the epitaxial Ni-Mn-Ga films on a Cr sacrificial layer [107] is presented. Fig. 6.1 gives an overview of the process flow for the sacrificial layer technology that contains the micromachining and the release of Ni-Mn-Ga films from a MgO substrate.

The process steps are summarized as follows:

1. Ni-Mn-Ga films are deposited on a single crystalline MgO (100) substrate by DC magnetron sputtering. The typical Ni-Mn-Ga film thicknesses range from 0.5 to 5 μm, which can be adjusted by the sputtering time. The distance between the target and substrate is maintained at 90 mm, the operating pressure is optimised at 0.008 mbar and the sputtering power is 100 W. The sputtering target is prepared by casting. Since the martensitic transitions strongly depend on composition, composition adjustment is a real challenge as commonly no stoichiometric transfer from the target to substrate occurs. The substrate temperature is identified as an essential process parameter to enable epitaxial growth, because of an increased loss of Mn and Ga with increasing temperature due to low vaporization temperatures [108]. Indeed fine-tuning of film composition can be adjusted by varying the substrate temperature. An epitaxial Cr buffer layer of 100 nm on the MgO (100) substrate is used as a sacrificial layer to release the Ni-Mn-Ga films later on. The chemical composition of the Ni-Mn-Ga films is determined by EDX to be $Ni_{50.5}Mn_{30.5}Ga_{19}$ (at. %).

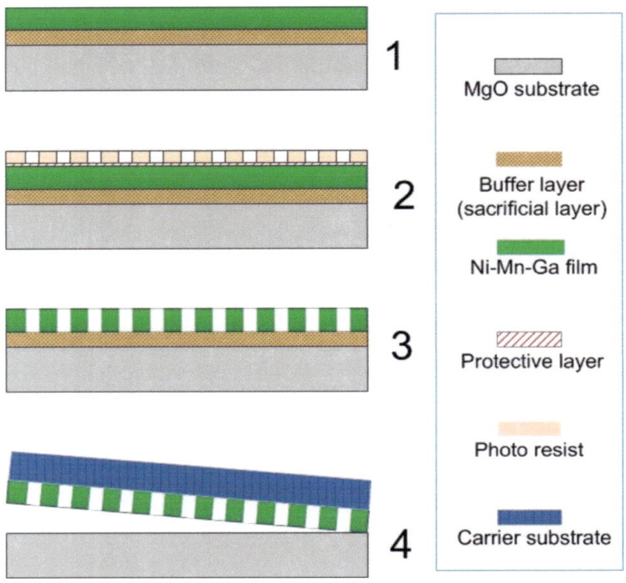


**Figure 6.1:** Process flow for fabricating a freestanding epitaxial Ni-Mn-Ga film based on the sacrificial layer technology.

2. Micromachining of the Ni-Mn-Ga films is performed by using optical photolithography. Before a metallic protecting layer of 100 nm is deposited on the sample. This protecting layer further helps to perform wet-chemical etching without damaging the film microstructure. Fig. 6.2 (a) shows Ni-Mn-Ga film stripe patterns after photolithography.
3. The selective wet-chemical etching is performed after photolithography. In this process step, the MgO substrate and Cr sacrificial layer have to fulfill the basic requirements of low roughness and high chemical stability. In addition, the sacrificial layer has to provide sufficient bonding strength during processing and to allow for selective removal. Ultimately, the sacrificial layer is etched out completely to release the micromachined Ni-Mn-Ga stripe patterns from the MgO substrate.

(b)

(a)

**Figure 6.2:** (a) The image shows stripe patterns after photolithography, (b) Freestanding Ni-Mn-Ga film stripe on an alumina substrate after transfer bonding technology.

4. Since the MgO substrate is not suitable for microsystems applications, the micromachined Ni-Mn-Ga film needs to be transferred to a target substrate. For this purpose, a novel transfer bonding technology is used [109], which has originally been developed for the batch integration of high quality materials like single crystalline silicon or shape memory alloys in microsystems applications. In transfer bonding, the micromachined target substrate consists of a receiver bonding site and an opening to allow for motion of the freely movable Ni-Mn-Ga film. First, it is bonded to the epitaxial Ni-Mn-Ga film before the sacrificial buffer layer is removed. Depending on the level of integration, it allows probing the freestanding films for studying intrinsic properties like crystallographic structure, magnetization measurements, mechanical properties, and the actuation performance respectively. Fig. 6.2 (b) shows a freestanding Ni-Mn-Ga film actuator bonded to an alumina substrate after transfer bonding.

After releasing the Ni-Mn-Ga films from the MgO substrate, first the properties of substrate-constrained and freestanding epitaxial Ni-Mn-Ga films are compared to evaluate the effect of this process on the film properties. Backen et al [110] investigated the structure and microstructure as well as magnetic properties before and after releasing Ni-Mn-Ga films from the MgO substrate. After deposition of Ni-Mn-Ga films on MgO covered by a Cr sacrificial layer, Auger electron spectroscopy (AES) results confirm no significant interdiffusion of Cr into the Ni-Mn-Ga film. In addition, Cr transfers the epitaxial relationship from the substrate to the Ni-Mn-Ga film due to a low lattice mismatch of 1.2 % of austenite Ni-Mn-Ga to Cr. Thus, Cr satisfies the key requirements as a suitable sacrificial layer. In addition, the microstructure of both, substrate-constrained and freestanding film having 1 μm thick is analyzed by means of SEM as well as AFM to evaluate the effect of etching solution on the surface twin morphology.

The constrained film is very flat on a length scale of several micrometers. In the sub-micrometer scale, one can see fine twins, which are rotated by 45° towards the substrate edges as shown in Fig. 6.4. This is an indication for the formation of 14M-layer modulated martensite [80]. By analyzing the height profile of twin morphology, the c/a ratio can be calculated. The geometrical model shown in Fig. 6.3 is used to calculate the c/a ratio by a simple expression

$$\frac{c}{a} = \tan^{-1}(45 - \alpha) \tag{0.1}$$

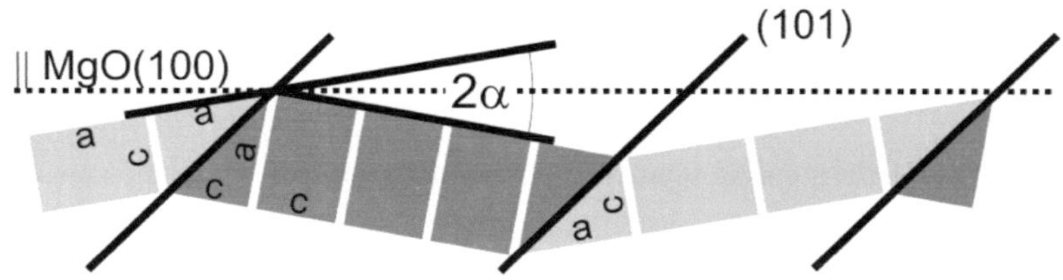


**Figure 6.3:** Geometrical model used to calculate the twin surface morphology and also used to calculate the c/a ratio from the height profiles measured by AFM.

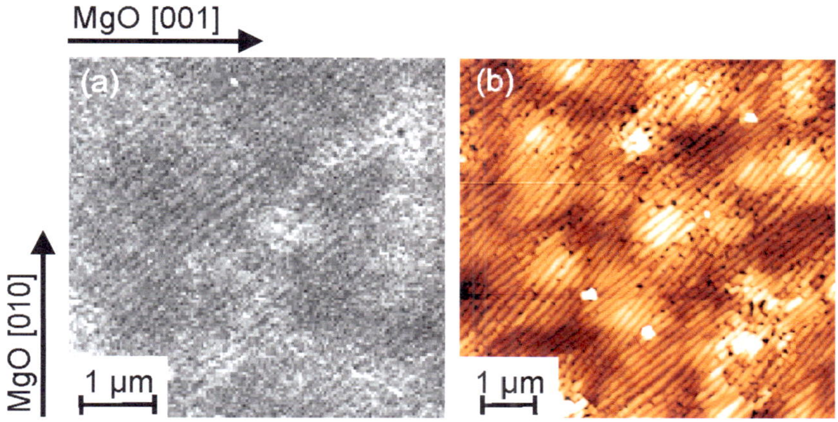


**Figure 6.4:** SEM (a) and AFM (b) micrograph of constrained film show flat and finely twinned microstructure. The AFM picture has a twin morphology height of 5 nm.

The analysis of AFM height profiles gives a c/a-ratio of about 0.9, which is another direct indication for 14M martensite [110]. SEM micrographs of the freestanding film reveal a wavy surface on a large scale as shown in Fig. 6.5(a). Due to releasing the film from substrate constraint, the film can expand in all directions, which leads to the waviness of the freestanding film. Looking at the length scale of a few micrometers as shown in Fig. 6.5 (b)-(d), the freestanding film reveals a finely twinned microstructure which is comparable to the constrained film. Both, the former surface (Fig. 6.5 (b)) and the former interface to the Cr buffer (Fig. 6.5 (c)) are finely twinned, which proves that the etching process does not affect the microstructure.

The c/a ratio of 0.9 of the freestanding film obtained from AFM height profile first confirms the non-destructive nature of the sacrificial layer technology and secondly indicates the formation of 14M martensite in the free standing films.

Variant reorientation in bulk Ni-Mn-Ga alloys has already been reported by many groups [24, 111, 112]. It has been shown that variant reorientation can be possible either by applying a magnetic field or external stress. Magnetization measurements are performed for constrained and freestanding films, but the measurements reveals no jumps in the magnetization indicating MIR [110]. This could be due to the microstructure as well as the coexistence of 14M and NM martensite phases; in particular NM martensite phase exhibits a low magneto-crystalline anisotropy and high twinning stress.

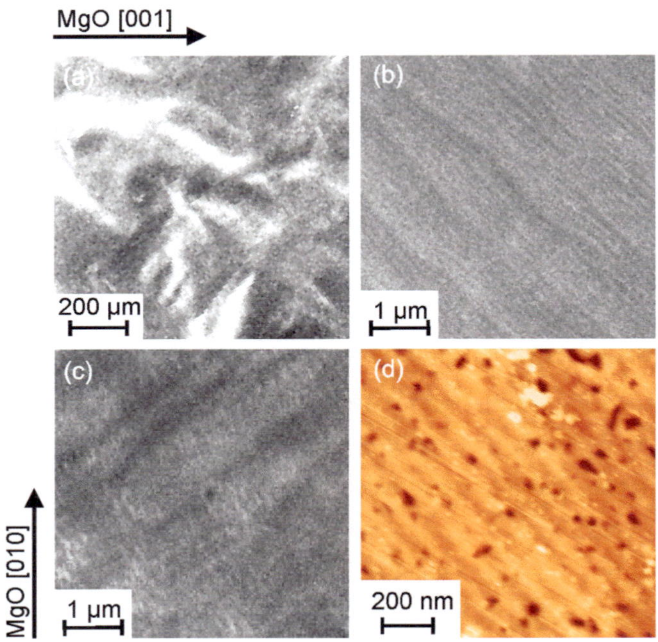


**Figure 6.5:** The surface morphology of freestanding Ni-Mn-Ga film. A wavy surface is observed over a length scale of several millimeters (a). SEM graphs of the former front (b) and backside (c) show finely twinned surface, which is confirmed by AFM (d). The height of twin morphology is 12 nm.

Nevertheless recently, Chernenko et al [113] reported large magnetic field induced strain of 0.17 % for NM martensite phase in bulk Ni-Mn-Ga alloys. However, the twinning stress and strain in the freestanding Ni-Mn-Ga films was not explored up to date due to technological problems in fabricating and handling freestanding epitaxial Ni-Mn-Ga films reported above. By this new technology presented here, the major technical difficulties have been overcome. This opens up the opportunity to investigate microstructure and mechanical properties of the freestanding epitaxial Ni-Mn-Ga films showing coexisting 14M and NM phases.

## 6.2 Microstructure and variant reorientation

For this investigation, freestanding Ni-Mn-Ga films are used after releasing from sample-1 type films. Before variant reorientation, the crystal structure, surface morphology and phase transformation characteristics are investigated to gain more insight about microstructure of freestanding Ni-Mn-Ga films in initial state. The direction-dependent mechanical properties are investigated. Then, after variant reorientation by a mechanical stress the microstructure of the freestanding films is evaluated.

### *6.2.1 Structural properties*

The freestanding epitaxial Ni-Mn-Ga films are analyzed by X-ray diffraction (XRD) at room temperature as shown in Fig. 6.6. The experiments on as-released freestanding films are performed in reflection mode at the ANKA synchrotron source with wavelength of 1.0332 Å. XRD characteristics of the freestanding films in the as-released conditions show strong reflections from (400) NM, (004) NM martensite as well as (400) 14M martensite and a weak reflection from (040) 14M martensite. However, high intense $Al_2O_3$ peaks are also visible due to freestanding film is supported at its both ends by $Al_2O_3$ substate, which can not be avoided from incident x-ray beam.

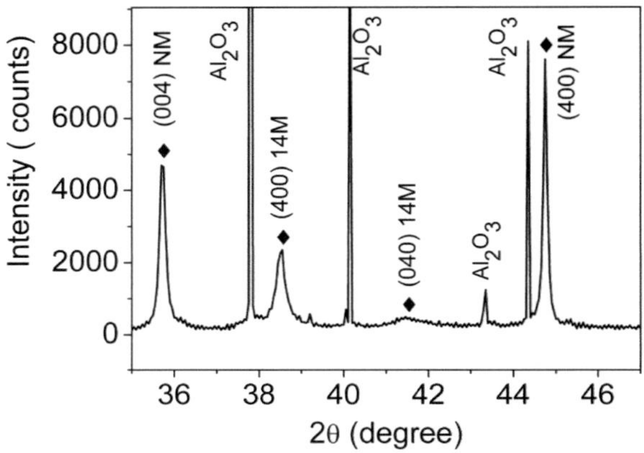


**Figure 6.6:** X-ray diffraction of Ni-Mn-Ga freestanding film (sample-1) supported by a polycrystalline $Al_2O_3$ substrate.

From the concept of adaptive martensite [84], it has been shown that 14M is a nanotwinned NM martensite phase and the microstructural transition between both phases proceeds by coarsening of twin boundaries. A key conclusion from these experiments is that films on a rigid substrate more easily form a modulated phase compared to bulk. For freestanding films, however one can expect a different behavior as neither stress-induced martensite helps to increase the martensitic transformation temperature nor the constraint of a rigid substrate hinders detwinning from 14M to NM.

### *6.2.2 Cross-sectional surface morphology*

Scanning electron microscopy (SEM) and FIB techniques are used to probe the surface and cross-sectional twin morphology of the freestanding epitaxial Ni-Mn-Ga films. The freestanding Ni-Mn-Ga films are cut along the cross-section and imaged with secondary electrons. The cross-section represents the (010) plane of Ni-Mn-Ga. Fig. 6.7 shows a cross-section of a freestanding Ni-Mn-Ga film prepared by FIB. The cutting edge is oriented along the [001] direction of the Ni-Mn-Ga unit cell. A wavy surface is observed exhibiting two different twin patterns. One pattern (type A) occurs at an angle of about 45° with respect to the surface edge with a periodicity of about 125 nm. This result is in line with AFM measurements reported in section 6.1 on the Ni-Mn-Ga films showing a curved height profile of similar periodicity. Based on the height profile, the c/a ratio of the film has been determined to be 0.9 corresponding to 14M martensite. In addition, a second pattern (type B) has been identified that is not clearly visible in Fig. 6.7 showing straight line features aligned perpendicular to the substrate edge [85].

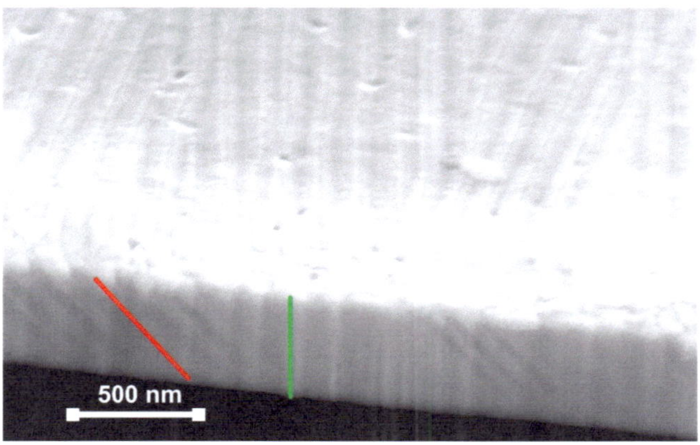


**Figure 6.7:** Surface and cross-section of a freestanding epitaxial Ni-Mn-Ga film in as-released state. The lines along the cross-section indicate the course of both types of twin patterns.

In cross-section, the two different twin patterns can be discriminated more clearly. The twin pattern of type A is inclined by 45° and of type B is aligned along the film growth direction. The periodicity of type B twins appears to be less regular than of type A varying from a few tens of nm to several hundred nm. Patterns exhibiting lines along the substrate edges have only been observed in films showing NM martensite by XRD [85] suggesting that these features are due to NM twin variants.

### *6.2.2 Thermally induced phase transformation properties*

The phase transformation temperatures of freestanding Ni-Mn-Ga stripes are investigated by four-probe electrical resistance measurements. Fig. 6.8 shows a typical electrical resistance characteristic of freestanding film.

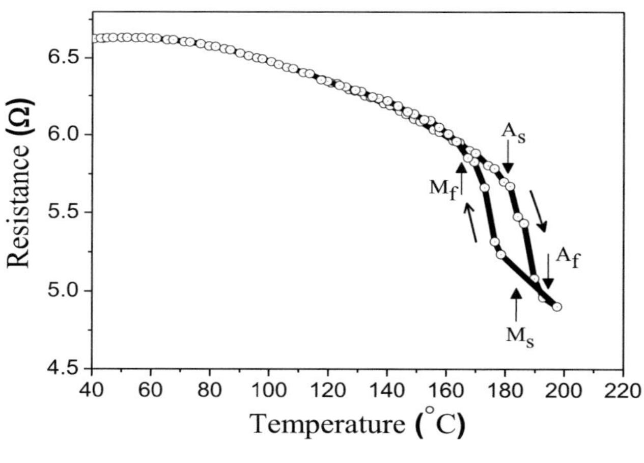


**Figure 6.8:** Electrical resistance characteristics of a freestanding Ni-Mn-Ga strip of 1 μm thickness along [100] Ni-Mn-Ga direction.

Upon heating, the austenite start temperature at $A_s$ = 182 °C and the corresponding finish temperature at $A_f$ = 191 °C is observed. Upon cooling, martensite formation starts at $M_s$ = 179 °C and ends at $M_f$ = 169 °C, respectively. It has been shown that the phase transformation temperatures of off-stoichiometric Ni-Mn-Ga alloys strongly depend on electron concentration [50, 114]. While for films on substrates often the stress between both increases the martensitic transformation temperature significantly [81], this can be excluded in the present case since no external stress is present in a freestanding film [14]. However, these results agree with results on NM martensite in bulk Ni-Mn-Ga single crystals [115]. The observed phase transformation temperatures are a characteristic feature of NM crystal structure, which is observed in X-ray diffraction as shown in Fig. 6.6.

### *6.2.3 Transmission Electron Microscopy (TEM)*

Fig. 6.9 shows a high angle annular dark field (HAADF)-scanning transmission electron microscopy (STEM) image of a Ni-Mn-Ga thin lamella in the as-released state. A fine twin pattern is observed varying in width from a few nm to about 65 nm, with an average width of 45 nm. All the twin boundaries in the cross section are aligned at an angle of ~45° with respect to the substrate normal. Throughout the film thickness, the twin width remains constant without branching. This is in contrast to films on rigid substrates where branching occurs to reduce the elastic energy at this incompatible interface [85]. As observed before by AFM, the bottom (top was partially eroded during deposition of protective layer) exhibits a facetted surface with angular differences of 9° between neighboring twins (Fig. 6.9 inset).

Fig. 6.10 shows a HR-TEM image of one of the twin interfaces in the as-released state seen in Fig. 6.9 above. The main lattice fringes seen on both sides of the interface are in basic agreement with the [010] zone of the 14M structure determined by XRD. However, with 3.19 Å for the (200) reflection and 2.75 Å for the (002) reflection, the lattice spacings measured in TEM differ slightly from the 3.12 Å and 2.80 Å estimated from XRD measurements (section 6.2.1). In addition, some larger scale variations are noticeable in the HR-TEM image, which are probably due to a Moiré pattern.

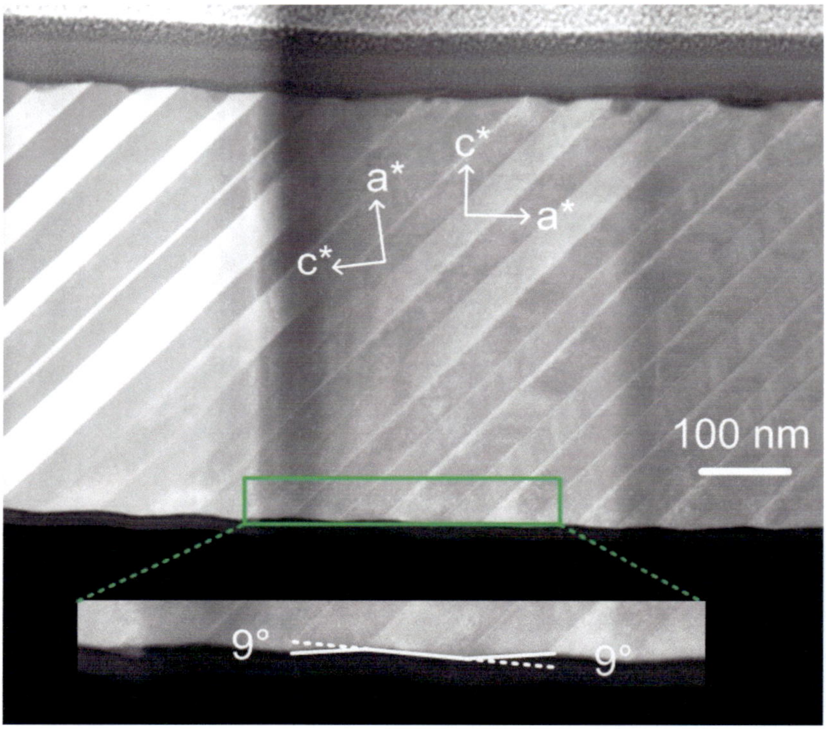


**Figure 6.9:** HAADF-STEM image of the mesoscopic twins in the Ni-Mn-Ga thin lamella in [010] orientation of Ni-Mn-Ga unit cell in the as-released state; the inset shows a bright field TEM image of the faceting of the twins at the bottom surface of the freestanding film.

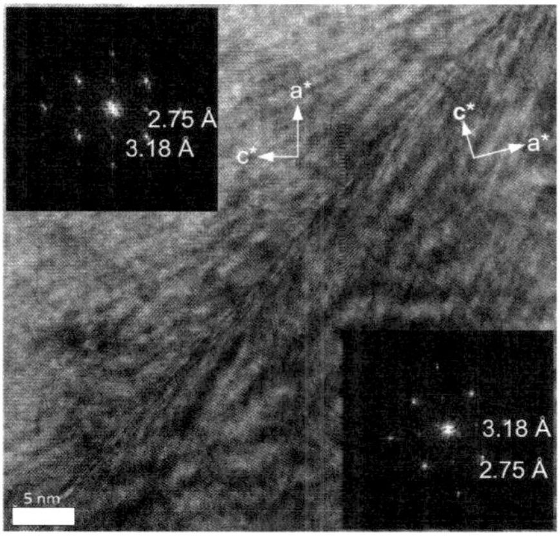


**Figure 6.10:** HR-TEM image of a twin boundary in the 14M Ni-Mn-Ga structure imaged along the [010] direction. The local FFT and the orientation of the crystallographic axes are shown as insets.

The direction of the crystallographic main axis is indicated in the HR-TEM image, which shows that the faceting of the twins corresponds to the (100) and (001) lattice planes of the 14M structure. Both planes are slightly inclined 4.5° to the substrate. The observed 9° angle for the faceting of the twins fits well to the orientation difference between the (100) and (001) lattice planes measured from the FFT of Fig. 6.10 (inset). Fig. 6.11 shows a selected area electron diffraction (SAED) pattern that confirms the presence of the twin structure of the 14M martensite variant imaged in [010] direction.

In addition, other reflections, with lattice spacings fitting to the NM martensite structure, are observed in the SAED pattern in agreement with the XRD results. This is probably due to a small amount of the NM structure projected on top of the 14M martensitic structure in the TEM lamella. Finally, some reflections at very low angles are observed, which are probably due to a multiple scattering of the 14M and the NM structure. However, further investigations are necessary to exclude that this corresponds to a larger super cell necessary to describe the 14M structure in full detail.

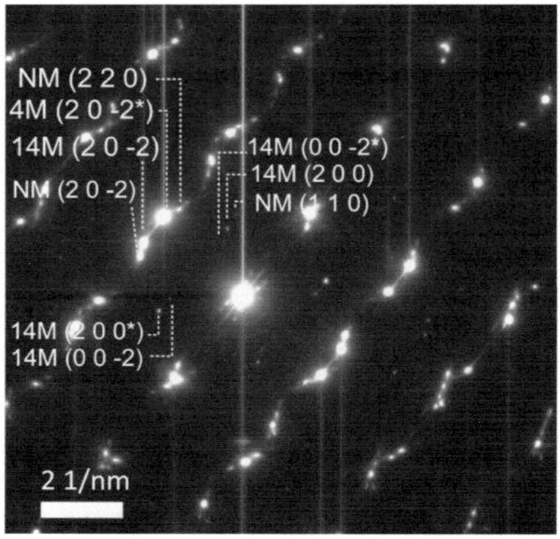


**Figure 6.11:** SAED pattern corresponding to an area with ~500 nm diameter. The two twin orientations of the 14M structure in [010] orientation can be clearly identified. In addition, some reflections are corresponding to the NM structure are also visible.

### *6.2.3 Tensile stress-strain characteristics*

The freestanding Ni-Mn-Ga stripes are investigated by tensile experiments in displacement-control mode using a strain rate of 0.01 mm/min. The tensile loading direction is applied in different directions with respect to the crystallographic direction of the films. The width and length of the Ni-Mn-Ga test specimens are 1 and 10 mm, respectively. Fig. 6.12(a) shows typical engineering stress-strain curves for a tensile load applied along [100] and [110] direction of the Ni-Mn-Ga film. In both cases, a broad strain plateau is observed after an initial elastic response. The schematic as shown in Fig. 6.12(b) reveals the direction of tensile stress acting on Ni-Mn-Ga unit cell.

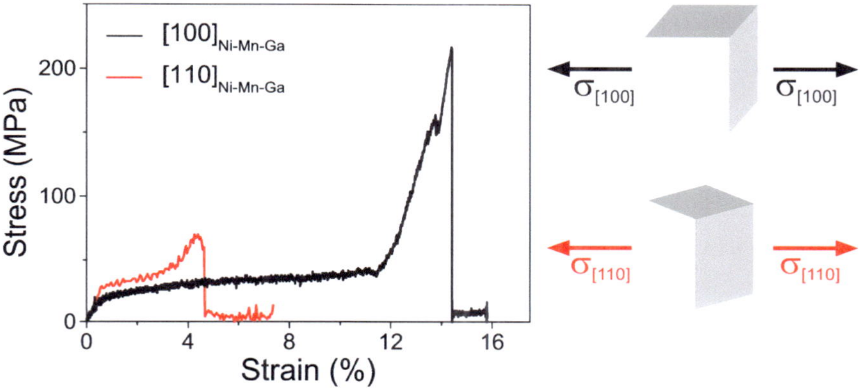


**(a)** **(b)**

**Figure 6.12:** (a) Engineering tensile stress-strain characteristics of freestanding epitaxial Ni-Mn-Ga stripes of 1 μm thickness along the [100] and [110] crystallographic direction of Ni-Mn-Ga as indicated, (b) Schematic representation of tensile stress direction acting on Ni-Mn-Ga unit cell

While in bulk often a step-like increase of stress level is observed during an intermartensitic transition from 14M to NM, this is not the case in the present experiments. This indicates that due to the coexistence of 14M and NM in the present samples no nucleation barrier seems to exist hindering the coarsening process from nanotwinned 14M to macrotwinned NM. The twinning stress is determined by the tangential method and is about 25 MPa for [100] orientation. Most strikingly, a large superplastic behaviour with a strain plateau of 12 % is observed, which indicates stress-induced reorientation of the NM martensite variants. Obviously, the films can hold very large stress values above 200 MPa showing no brittle behaviour like polycrystals. For [110] orientation, a different mechanical behavior is observed. In this case, a higher twinning stress of about 30 MPa is found and a considerably smaller strain plateau of 4 % occurs. As NM martensite exhibits a higher c/a ratio than 14M it can compensate more strain. In previous studies, the lattice constants of the NM structure have been determined to be a = b = 5.46 Å and c = 6.76 Å, which gives an upper limit of strain of |1-c/a| = 23% [110]. As the Ni-Mn-Ga stripes have not been subjected to mechanical loading or a magnetic field, one can assume that they are in a multi-variant state before straining.

For tensile loading in [100] direction, variants with the long c-axis orientation along the loading direction do not contribute. Thus, the maximum strain is given by the fraction of variants with a-axis orientation along the loading direction, which is estimated from the experiment to be about 50 %. For tensile loading in [110] direction, the maximum strain is expected to be lower since only the projection of the [100] direction onto the loading direction is relevant. The schematic shown in Fig. 6.13 illustrates the martensite variants contributing to strain along both (100) and (110) orientation of Ni-Mn-Ga unit cell.

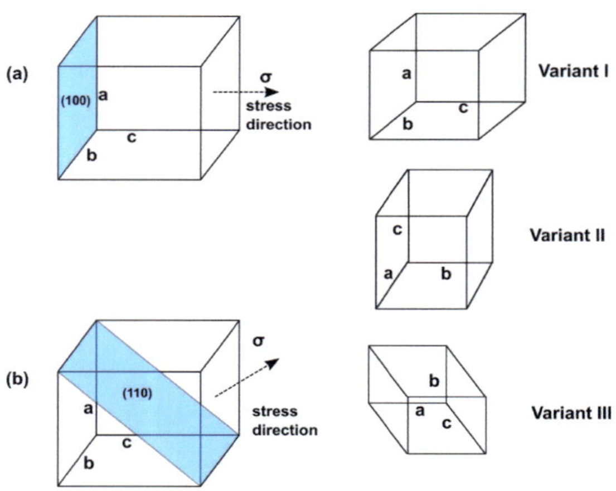


**Figure 6.13:** Schematic of the Ni-Mn-Ga unit cell subjected to tensile stress (a) along (100) direction, (b) along (110) direction. Three different martensite variants with lattice parameters are shown.

Further microstructural investigations on the fraction of contributing variants in initial state will be required for a quantitative understanding. The difference in twinning stress for tensile loading along the [100] and [110] directions of the Ni-Mn-Ga unit cell can be explained by analyzing the shearing stress which acts on the twin boundaries following the concept of Otsuka and Wayman [18]. The shear stress acting on the twinning plane is shown in Fig. 6.14(a), where F is the force acting on the twining plane, A is the area of cross section and N is the normalization vector. The stress necessary to shear two {110} type twin boundaries is equal to $\tau_{\{110\}} = \sigma \cdot S$ with σ being the external stress and the Schmid factor $S$ [116]. The Schmid factor is a geometrical factor and can be calculated from $S = \cos\lambda \cdot \cos\kappa$, with λ and κ being the angles between the external stress direction given by F and the shearing direction (τ) and the normal of the {110} plane (N), respectively. Consequently, σ is only dependent on the Schmid factor $\sigma = \tau_{\{110\}}/S$.

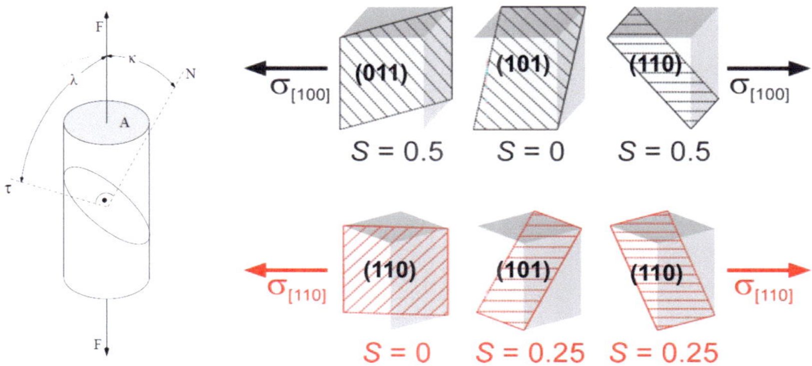


(a) (b)

**Figure: 6.14** (a) Schematic of the shear stress acting on a crystallographic plane, (b) possible orientations of {110} planes in a cubic unit cell and the Schmid factors S for both stress directions; legend: F = external stress direction, N = normalization vector, τ = shear stress, A = area of the cross-section.

The slide system consisting of {110} planes and the corresponding <110> shear direction with high values of $S$ are expected to be activated first whereas slide systems with $S = 0$ are not activated at all. The value of $S$ is calculated for the different {110} planes and their corresponding shearing direction for the case of a cubic unit cell (Fig. 6.14 (b)). The Schmid factor calculated for the external load in [100] and [110] direction is twice as high than along [110] direction. From this one would expect a factor of two difference between the twinning stress when loading in [100] and [110] directions. The experimentally observed difference is smaller (25 and 30 MPa). The tetragonal distortion of the martensite results in angular deviation from the cubic unit cell used here. In case of the [100] direction this results in a reduction of $S$ from the maximum value of 0.5, while for the [110] direction $S$ may increase for some tilt and rotation.

In conclusion, Ni-Mn-Ga stripes are orientated along the different crystallographic directions allowing direction-dependent tensile measurements after substrate release. X-ray diffraction results on the freestanding films show major contributions of NM and minor contributions of 14M martensite. The NM structure is also reflected in the high phase transformation temperatures observed by electrical resistance measurements. For the [100] and [110] directions, the tensile measurements reveal superplastic behavior with a strain 12 % and 4 %, respectively. The orientation dependence of twinning stress can be qualitatively explained by the different orientations of external tensile load with respect to the twinning planes.

### *6.2.4 Structure after variant reorientation*

Within the Ni-Mn-Ga system various modulated (10M and 14M) and non-modulated martensite phases are observed. From the concept of adaptive martensite, the transition from 14M to NM martensite is considered as a microstructural transition proceeding by annihilation of nanotwin boundaries [85]. The collective defect energy associated with the fractal process of coarsening is interpreted as an energy barrier separating the 14M and NM microstructure. This differs from the usual approach, which considers this transformation to be a structural intermartensitic transition. However, when freestanding films exhibiting mixed 14M and NM martensitic variants are subjected to variant reorientation either by mechanical stress or by magnetic field the final resulting martensite can have two possible selections of martensite variants. First, after variant reorientation 14M single variant and NM single variant states can be formed. Second, the intermartensitic transformation between 14M and NM is possible and it form NM single variant state. Usually either one is expected to occur in freestanding films after variant reorientation.

Stress-strain experiments are helpful to clarify the different viewpoints as explained in Fig. 6.15..

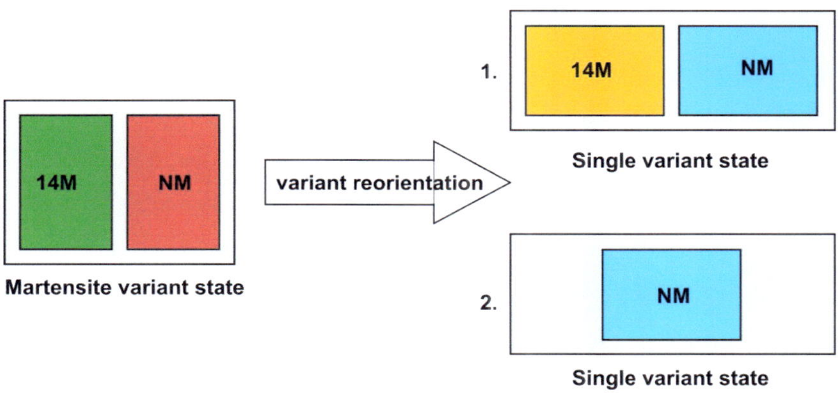


**Figure 6.15:** The schematic explains two different possible selections of martensite variants upon stress induced variant reorientation of mixed 14M and NM martensites.

In bulk a step-like increase in stress level is often observed [87]. This is commonly considered as an intermartensitic transformation towards a martensite phase exhibiting a higher tetragonality. At low stress levels variant reorientation within a low-tetragonality phase occurs. When this is no longer possible, a transformation towards a phase with higher tetragonality still allows stress to be a accommodated by variant reorientation. Thin films can only be loaded in tensile mode in order to avoid buckling. It is essential to explore crystal structure and microstructure after tensile straining to gain insight in the stability of modulated phases and the effect of mechanical training to improve twin boundary mobility. Here, freestanding epitaxially grown Ni-Mn-Ga films with coexisting phases of 14M and NM martensite are investigated. These films exhibit a large anisotropy of stress and strain including a pronounced superplastic behavior in tensile tests (Section 6.2.3). X-ray diffraction, focused ion beam (FIB) and transmission electron microscopy (TEM) measurements are presented after tensile loading induced superplastic straining.

The structure of the freestanding epitaxial Ni-Mn-Ga films is analyzed by X-ray diffraction at room temperature. The experiments on strained films are performed in transmission mode at the ANKA synchrotron source with wavelength of 1.0332 Å. XRD characteristic of the freestanding films in the as-released condition is shown in Fig. 6.6, which reveals strong reflections from (400) NM, (004) NM martensite as well as (400) 14M martensite and a weak reflection from (040) 14M martensite. The freestanding epitaxial Ni-Mn-Ga stripes are subjected to tensile loading along $[100]_{Ni\text{-}Mn\text{-}Ga}$ direction (Fig. 6.12) to investigate the influence of stress-induced reorientation on the film structure. Fig. 6.16 shows an XRD measurement of a freestanding Ni-Mn-Ga stripe after tensile loading to a strain of 12 %. In this case, a single diffraction peak due to (400) NM martensite is observed indicating that the film is in a single variant state of NM martensite.

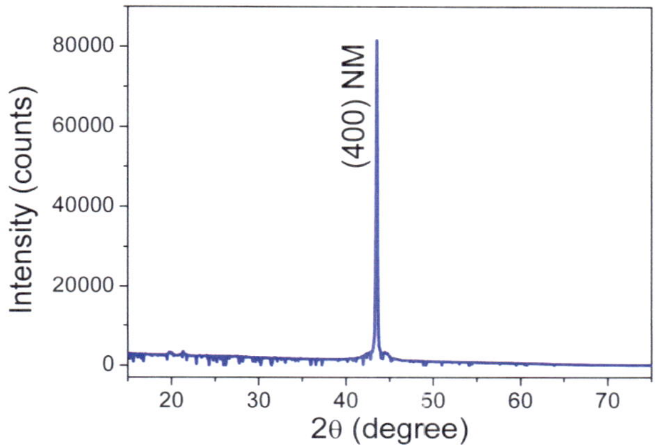


**Figure 6.16:** X-ray diffraction measurement of the freestanding Ni-Mn-Ga film (sample-1) stripe after tensile loading to a strain of 12 %.

The peak position is slightly shifted with respect to the as-released state in Fig. 6.6, which might be attributed to the high strain level. In the measurement conditions, lattice planes parallel to the film surface are probed. Since [400] is the short axis, the tensile in-plane stress apparently aligns the long [004] axis along the stress direction. Obviously, stress-induced variant reorientation caused by tensile loading is accompanied by a complete transition to NM martensite. NM martensite exhibits a higher *c/a* ratio than 14M martensite and can hence compensate more strain.

Scanning electron microscopy (SEM) and FIB techniques are used to probe the surface and cross-sectional twin morphology of the freestanding epitaxial Ni-Mn-Ga films and is shown in Fig. 6.17. The cross-section represents the (010) plane of Ni-Mn-Ga. The investigations are carried out in superplastically strained condition after complete variant reorientation. In this case, a flat and homogeneous surface is observed that does not show any traces of twin patterns neither on the surface nor in cross-section. Only small holes are visible on the surface. The holes partly result from the etching process during micromachining of the Ni-Mn-Ga stripes. In addition, accumulation of defects at the surface during variant reorientation due to stress localization cannot be excluded [117].

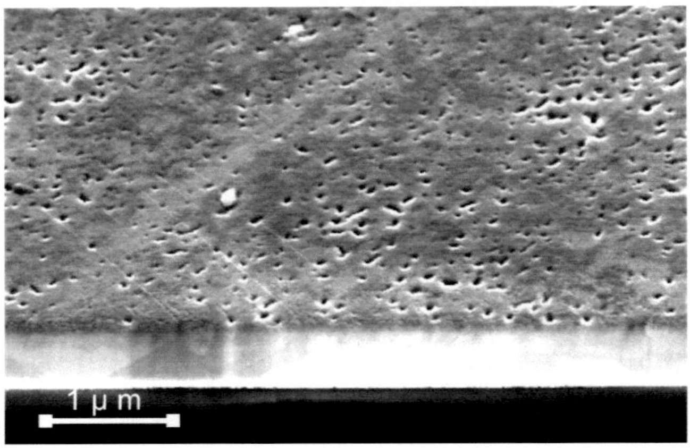


**Figure 6.17:** Surface and cross-section of a freestanding epitaxial Ni-Mn-Ga film after variant reorientation as prepared by FIB.

The disappearance of type A twins (Fig. 6.7) is in line with the disappearance of 14M reflections in the XRD measurement shown in Fig. 6.16. The resulting microstructure is thus attributed to single variant of NM martensite. The disappearance of the modulated microstructure can be explained by the concept of adaptive martensite. In the adaptive concept, there are several generations of twinning that have to be taken into consideration (Section 2.5.1). The highly mobile twin boundaries of the 14M structure represent the second generation of twinning. The 14M structure itself is built of tetragonal nano twins of the non-modulated martensite which represent the first generation of twinning. When the 14M structure is subjected to a tensile load along [100] direction, detwinning occurs. This means that the mobile twin boundaries proceed through the crystal resulting in a single 14M variant. With further increase of the tensile stress, the 14M structure is coarsened. The result is a single variant of non-modulated martensite with its long c-axis aligned along the load direction. In our experiment, a stress level of 25 MPa was enough to fully transform 14M to a single variant of non-modulated martensite. Fig. 6.18 shows a bright field TEM image of a Ni-Mn-Ga lamella prepared after stress-induced variant reorientation. In this case, no features of twin boundaries are observed along the cross-section in agreement with SEM-FIB results of detwinned NM martensite.

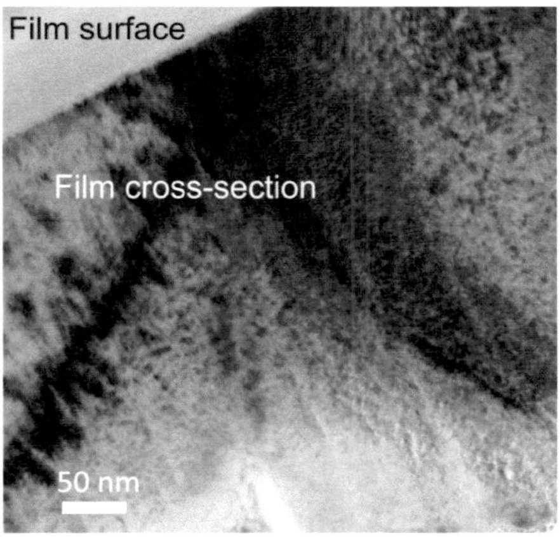


**Figure 6.18:** Bright field image of Ni-Mn-Ga TEM lamella after variant reorientation.

In conclusion, the microstructure of the freestanding epitaxial Ni-Mn-Ga films is investigated by FIB and TEM measurements showing coexisting phases of 14M and NM martensite after release from MgO (100) substrate. In the as-released state, SEM-FIB cross-sections reveal two different twin patterns on a mesoscopic scale, a first twin pattern being inclined by 45° and a second twin pattern parallel to the film growth direction. TEM investigations identify the twin boundaries aligned 45° with respect to the film surface as mesoscopic 14M twins.

After superplastic straining the film by 12%, no twin patterns are observed. The film is in a single variant state of NM martensite. Tensile loading of 25 MPa appears to be sufficient for detwinning the 14M structure, which is composed of nanotwinned NM variants. The observed effect of tensile loading on phase formation has to be taken into account for the development of a training process to reduce the twinning stress and to enable magnetic-field induced reorientation. Any training process should aim at supporting the formation of mesoscopic twin boundaries showing macroscopically observable high mobility. Our results indicate that pure mechanical training will not be effective. Instead, the combinations of thermo-mechanical or thermo-magneto-mechanical training are suggested as a route to achieve this goal.

## 6.3 Transformation behavior

The transformation behavior of epitaxial Ni-Mn-Ga films is analyzed by combining temperature dependent measurements of electrical resistance, X-ray diffraction (XRD) and tensile stress-strain characteristics. For this investigation, Ni-Mn-Ga thin films (samples of type 2) deposited at 300°C are considered. The sputtering power of 70W is applied to deposit films. By means of energy-dispersive X-ray spectroscopy (EDX), the film composition is determined to be $Ni_{47.7}Mn_{31.1}Ga_{21.2}$. Decrease of deposition temperature from 400°C to 300°C significantly yield increase in Mn and Ga compositions. The as-deposited Ni-Mn-Ga films are machined to lamellar stripe specimens by optical lithography and wet-chemical etching. The edges of the stripe pattern are oriented along the [100] direction of the Ni-Mn-Ga unit cell, which corresponds to an angle of 45 degree with respect to the MgO <100> substrate edges.

### *6.3.1 Surface morphology*

The surface morphology of freestanding epitaxial Ni-Mn-Ga films is studied by Scanning Electron Microscopy (SEM-Jeol) with back-scattered electron contrast. The surface morphology of a freestanding epitaxial Ni-Mn-Ga stripe is shown in Fig. 6.19. The traces of twin boundaries which are aligned parallel or perpendicular to the <100> directions of the austenite Ni-Mn-Ga unit cell are observed. The similar surface morphology has been reported for the constrained Ni-Mn-Ga films on a MgO substrate [80]. Since the twin boundaries are connected by martensite variants with a- or c-axes being oriented in out-of-plane direction, the morphology feature of sample of type 2 is quite different than sample of type1 due to different deposition temperature and process parameters that are maintained during sputtering. Hence the resulting alignment of twin surface morphology differs from Fig. 6.7.

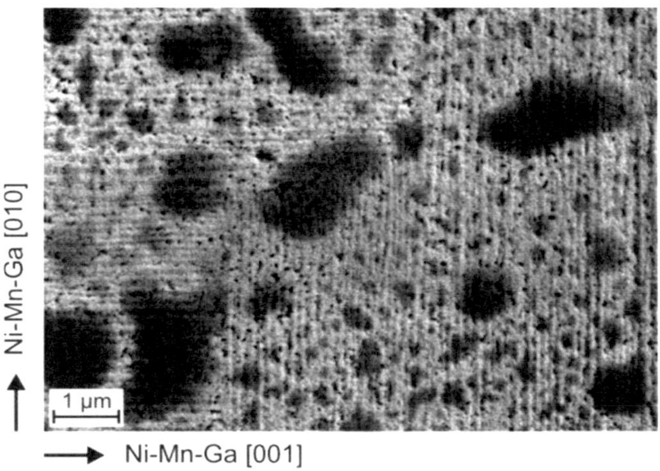


**Figure 6.19:** SEM image of a freestanding epitaxial Ni-Mn-Ga film showing traces of fine twin boundaries along the <100> directions of the austenitic Ni-Mn-Ga unit cell.

### *6.3.2 Magnetic properties*

Temperature-dependent magnetization characteristics of the Ni-Mn-Ga film shown in Fig. 6.20 are determined using a Vibrating Sample Magnetometer (VSM, Quantum Design PPMS). When cooling the sample from 120 °C, an increase of magnetization occurs at about 90 °C, which can be attributed to the transition from paramagnetic to ferromagnetic state. When the temperature is further decreased, a step-like drop of magnetization is observed. Commonly, this drop in magnetization is explained by a transformation from austenitic phase with low magnetocrystalline anisotropy to martensite having a strong uniaxial anisotropy [80]. Upon heating the sample, the reverse transformation is observed. The corresponding phase transformation temperatures are determined by the tangential method. The martensitic start and finish transformation temperatures $M_s$ and $M_f$ are determined to be 62 and 49 °C, respectively. The start and finish transformation temperatures of the reverse transformation $A_s$ and $A_f$ are 53 and 70 °C, respectively.

The inset in Fig. 6.20 shows a magnetic hysteresis loop measured at room temperature. The jump of magnetization at about 37 mT indicates a partial reorientation of martensitic variants. The jumps in the hysteresis loop have been analyzed in detail by Thomas et al [80]. It was shown that it originates from an internal reorientation within the 14M martensite and it can be possible even in films constrained by a rigid substrate. When applying a magnetic field, the fraction of variants with b-axis along the field direction increases on cost of the other two variant fractions. Since this configuration is only stable in high magnetic fields, the mechanism is reversed when reducing the external magnetic field. Therefore, the jump in magnetization has to be observed upon increasing and decreasing the magnetic field.

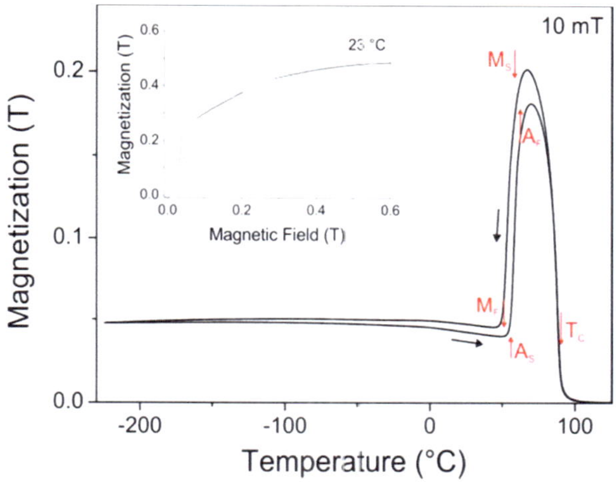


**Figure 6.20:** Temperature-dependent magnetization curve of the Ni-Mn-Ga film on the MgO (100) substrate showing a transformation from austenite to martensite and the ferromagnetic transition at Curie temperature $T_c$ as indicated. The characteristic jump in the magnetization curve measured at room temperature (inset) indicates magnetically induced variant reorientation. In both cases, the magnetic field is applied along the [001] direction of the Ni-Mn-Ga unit cell.

These results may be interpreted in terms of a transition from a phase with lower magnetocrystalline anisotropy to one with a higher anisotropy taking place in the temperature range between 49 and 62 °C. At room temperature, the constrained film shows traces of the fine twin boundaries parallel to <100> direction of Ni-Mn-Ga as shown in Fig.6.19. These 14M twin boundaries can be moved by an external magnetic field that leads to a partial variant reorientation as shown in Fig. 6.20 (inset).

### *6.3.3 Electrical resistance*

As a next step, the film has been released from the substrate in order to enable free elongation. The transformation behavior of the freestanding Ni-Mn-Ga film is probed by electrical resistance measurements as shown in Fig. 6.21.

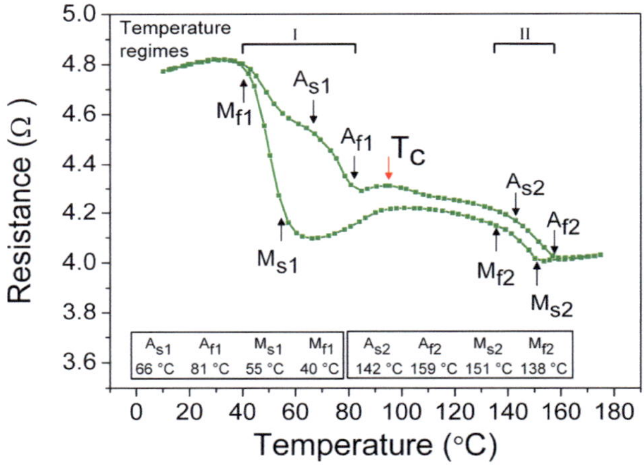


**Figure 6.21:** The freestanding Ni-Mn-Ga film shows a two-stage transformation behavior upon cooling and heating probed by electrical resistance measurement.

The measurement reveals a two-stage transformation behavior of the freestanding film. After an initial drop of resistance during cooling, the resistance increases at about 150 °C. The slope of the curve changes gradually down to 90 °C, where the resistance drops again. A second, sudden increase of resistance can be observed between 60 and 40 °C. With further cooling the sample, the resistance decreases again. Upon heating, a large hysteresis can be observed in the range between 40 and 80 °C in the temperature regime (I) as indicated in Fig. 6.21. The kink in the curve at about 90 °C is also observed in the heating cycle. Further heating up to 155 °C leads to a gradual decrease of resistance, afterwards it increases again. The electrical resistance measurement shows more complex transformation behaviour than the magnetization measurement performed on the same sample on a MgO substrate, which is not sensitive to the temperature regime above the Curie temperature. The electrical resistance measurements indicate that further transformations occur above $T_C$ and it is indicated as transformation regime (II). This is in clear contrast to previous interpretations of magnetization measurements revealing the presence of martensite transformation at higher temperature.

Magnetization measurements are indirect, which allows considering e.g, an intermartensitic transition between 14M and NM as an alternative origin of the observed change of low field magnetization in the temperature regime (I). Also for this transformation the magnetocrystalline anisotropy increases from 14M to NM, which may explain the observed drop of magnetization at lower temperatures. The thermal hysteresis observed in the magnetization measurement corresponds to the one observed in the low temperature regime (I) in the electrical resistance measurement. At the first glance this indicates that this transformation occurs between austenite and 14M martensite. Thus, transformation regime (II) in resistance measurements could be due to transformation between NM martensite and austenite.

### *6.3.4 X- Ray diffraction*

In order to investigate the structure of the freestanding film, XRD measurements are performed at different temperatures, while cooling a freestanding Ni-Mn-Ga film from 200 °C down to room temperature. Temperature-dependent X-ray diffraction measurements are carried out at the Angstroemquelle Karlsruhe (ANKA) at the wavelength of 1.504 Å. The temperature is varied between 25 and 200 °C. The measurements are performed on a freestanding Ni-Mn-Ga film. The X-ray diffraction measurements are performed at three different temperatures upon cooling from high temperature (200°C). In contrast to previous measurements performed at room temperature, this temperature dependent measurement allows probing martensite structure at different transformation regimes.

Fig. 6.22 shows XRD diffractograms of a freestanding Ni-Mn-Ga film at three different temperatures. Only {004} type reflections are observed, reflecting the epitaxial growth of the Ni-Mn-Ga film. At 145 °C, one major peak is observed at 61 degrees that is attributed to the B2 structure of austenite. In vicinity of this major reflection several minor reflections are observed. The small peak splitting hints to a martensitic phase with low distortion, which allows speculating on a premartensite with increase of tetragonal distortion.

When the temperature decreases further down to 120 °C, the peak at 61 degrees decreases, while additional peaks occur at 65.5 degrees that can be assigned to (400) NM martensite. An unambiguous indexing of the diffraction peak at 61 degrees is not possible, as the (040) 14M peak is expected to occur very close to the (400) A peak. At 55 °C, the number of observed peaks is even richer. Reflections of both, 14M and NM martensite, are observed, but even a coexistence of both phases does not allow a complete indexing of all reflections.

The XRD measurements prove the presence of martensite well above $T_c$, but clear identification of the different structures cannot be given. The reduced peak splitting at higher temperatures suggests that the martensitic distortion is reduced at higher temperatures. However, these measurements clearly show that changes in (micro-) structure even occur between $M_{s2}$ and $M_{s1}$. Together with the low hysteresis in temperature regime (II) and curved change of resistivity between temperature regime (II) and $T_c$, it is possible to speculate that the high transformation is of weakly first order without an abrupt change of lattice parameters.

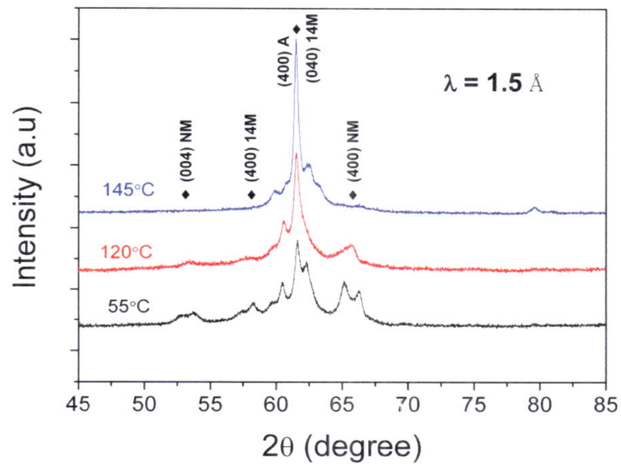


**Figure 6.22:** Temperature dependent X-ray diffraction characteristics of a freestanding Ni-Mn-Ga film after cooling from high temperature (200 °C).

Since also in bulk materials often an increase of tetragonal distortion with reduced temperature has been reported [99], the following scenario is suggested, which is based on the concept of adaptive martensite. The low tetragonal distortion at high temperatures results in a low twin boundary angle. As the twin boundary energy commonly increases squared with tetragonal distortion [118] at high temperatures an adaptive martensite can form. When reducing temperature and thus increasing tetragonal distorsion, twin boundary energy is expected to increase. At a certain temperature coarsening should occur, which reduces the twin boundary energy by transforming an adaptive microstructure towards a tetragonal NM martensite [85]. Furthermore intermartensitic transformations are possible, which occur between martensites exhibiting different distortions. Driven by a reversible change of tetragonal distortion these intermartensitic transformations can be reversible, but exhibit a hysteresis since the (adaptive) microstructures do not allow a direct transformation between two structures of the same crystal symmetry. It is suggested that such type of intermartensitic transformation occurs in temperature regime (I). However, it is still unclear about the contribution of austenite phase above $T_c$ from XRD measurements.

### *6.3.5 Mechanical properties*

The mechanical properties of the Ni-Mn-Ga stripes are investigated in a tensile testing machine (INSTRON) in strain control mode. Engineering stress–strain characteristics are determined for as-released freestanding Ni-Mn-Ga stripes at different ambient temperatures. For mounting the stripe specimens, a supporting frame is used to avoid any pre-stress. The strain is applied with a strain rate of 0.009 mm/min in uniaxial direction of the Ni-Mn-Ga stripe, which coincides with the [001] direction of the Ni–Mn–Ga unit cell.

The mechanical properties of freestanding epitaxial Ni-Mn-Ga stripes are investigated by tensile experiments along the [001] crystallographic direction of the Ni-Mn-Ga unit cell. Fig. 6.23 shows engineering stress-strain characteristics performed at different temperatures. Each experiment was performed on a new stripe made of the same film, hence each measurement started at the same multivariant state present after film release.

At room temperature, a broad stress plateau up to a strain of about 14 % is observed. In the stress plateau, the variation of stress is rather smooth showing only small stress drops indicating that the transformation is homogeneous. Further straining results in a rapid increase of stress. The experiment is stopped at a stress of 40 MPa to avoid overloading. The twinning stress determined by the tangential method is about 22 MPa.

At 70 °C, the stress plateau occurs at a lower value of 15 MPa. In this case, the maximum strain is reduced to about 12 %. At this temperature we observe a substantial increase of noise, which will be discussed later. Nevertheless, a number of remarkable features can be noticed. At low strain, a rather gradual increase of stress occurs followed by a significant stress drop at about 9 MPa indicating reorientation in a larger section of the stripe specimen. At larger strain, the stress increases further up to 15 MPa until the final stress plateau is reached. In the stress plateau, the stress-strain characteristic shows further relatively large stress drops.

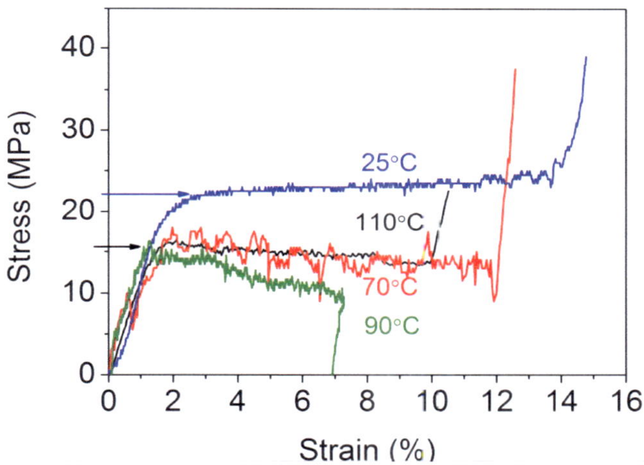


**Figure 6.23:** Tensile stress-strain characteristics of freestanding Ni-Mn-Ga stripes of 2 μm thickness at 25 °C, 70 °C, 90°C and 110 °C.

At 90°C, the onset stress for the plastic deformation is 15 MPa. Increase in deformation strain results in the slight decrease of stress due to temperature drop from 90°C to 86°C between start and end of the experiment. The experiment is stopped at 7% plastic strain to analyze the unloading path of stress-strain characteristic. Interestingly, the unloading characteristic does not reveal reversible pseudo elastic deformation. This confirms the martensite state of the tensile specimen and it excludes the presence of austenite at this operating temperature. However, the small stress drops in the stress-strain characteristics can be distinguished due to variant reorientation.

At 110 °C, the stress-strain characteristic shows a more homogeneous behaviour similar to room temperature. The stress plateau is reached in a single step. The initial slope is steeper compared to the case of 70 °C, but less steep compared to the measurement at room temperature. The stress plateau observed at 110 °C corresponds to the second stress plateau observed at 70 °C. It is rather smooth showing only small stress drops. Even at these high temperatures no pseudoelastic behaviour is observed, which excludes the presence of a large fraction of austenite. The observed decrease of twinning stress with increased temperature agrees with bulk experiments [99]. The stress-strain measurements confirm the complex transformation behavior observed in four-probe resistance and X-ray diffraction measurements. First, a strain above 10 % at all temperatures is observed. Considering that the starting variant distribution is the as-released multivariant state [119], this proves that the ground state is a detwinned NM martensite. Second, a significant increase of the twinning stress below temperature regime (I) is observed, which agrees well with the analysis of Straka et al [99] which imply that intermartensitic transformations have a large impact on twinning stress. Third, stress-strain measurements at 70° and 90° are substantially noisier compared to all other temperatures. Since this temperature is within the hysteresis of transformation (temperature regime (I)) an easy switching between both degenerated states should be possible. In addition to thermal fluctuations, the released transformation enthalpy should result in a discontinuous movement of the transformation front. Hence, stress-strain measurements are more indicative than XRD measurements. One can attribute the observed richness of XRD reflections, which cannot simply be assigned to A, 10M, 14M or NM, to the adaptive origin of modulated phases. According to the adaptive concept by Khachaturyan et al [120] the geometrical constraint at the habit plane induces the modulation, which are interpreted as a periodic arrangement of twin boundaries at the nanoscale. While for fixed films on the substrate constraint supports the formation of adaptive phases [85], in thin freestanding films this constraint is absent. Therefore the habit plane is expected to change from a two dimensional plane to a one dimensional line [121]. This constraint, however, defines the twin stacking sequence. As analyzed by Ustinov et al [122], already slight variations in stacking sequence result in substantially different superstructure diffraction patterns. Further theoretical analysis of this complex diffraction pattern may allow understanding the changes in constraints in more detail.

Finally, other alternative explanation needs to be considered. For instance the sample may be inhomogeneous and e.g., 14M and NM transform separately in the temperature regimes (I) and (II), respectively. Depending on composition (*e/a*-ratio) different martensitic phases like 14M and NM occur, which exhibit specific transformation temperatures and twinning stress as shown for single crystals [115]. Despite their noisiness the stress-strain measurements at 70° give indications for a two-phase behavior. However, one can consider this explanation as unlikely since our previous TEM analysis of freestanding film did not reveal any inhomogeneities on large or small scale, respectively [119]. In addition the stress-strain behaviour observed at room temperature and at 110 °C between regimes (I) and (II) showing homogenous reorientation behaviour despite the coexistence of 14M and NM martensite (Fig. 6.23). Furthermore, no pseudoelastic behaviour is observed even at 110 °C, which excludes a large fraction of austenite between temperature regimes (I) and (II).

In conclusion, a complex transformation behavior is formed in epitaxial Ni-Mn-Ga films. A two-stage transformation in the temperature ranges (I) and (II) is observed. The combination of experiments allows excluding one simple austenite-martensite transformation. It is suggested that instead a weakly first order transformation without an abrupt change of lattice parameters occurs at temperature regime (II), followed by an intermartensitic transformation in temperature regime (I).

## 6.4 Training of freestanding Ni-Mn-Ga films

Bulk Ni-Mn-Ga alloys are subjected to thermo-mechanical training in order to decrease the twinning stress and improve the twin boundary mobility. Thermo-mechanical training decreases the onset stress needed for the martensite variant reorientation [123]. In thermally formed self-accommodated martensite the orientation of the martensite variants are usually random, i.e., in multi variant state. Thermo-mechanical treatment may be applied in order to obtain a variant structure with a desired texture. When the material is stressed during cooling over the martensite temperature regime, the low-symmetry martensite phase orients to accommodate the stress in such a way that a predominant variant is formed. For the 10M or 14M tetragonal martensite with $c/a<1$ compressive stress is more applicable, while in the NM martensite with $c/a>1$ tensile stress is more effective.

Thermo-mechanical training is performed on freestanding Ni-Mn-Ga samples of type 3, which are sputter deposited on MgO (100) substrate with 150 nm Cr as a sacrificial layer (see Table 6.1). The sputtering power of 100W is applied to deposit the films. Deposition is performed at 350°C resulting in 2 µm thick films with $Ni_{49.2}Mn_{32.4}Ga_{18.4}$ composition. The e/a ratio of these samples is calculated to be 7.74. The samples of type 3 are thicker than sample of type 1, their chemical composition deviates particularly from sample of type 2. Using photolithography, the as-deposited films are machined to lamellar stripes of 10 mm and 1 mm and released from the substrate by wet-chemical etching. The edge of the stripe pattern is oriented along the [100] direction of the Ni-Mn-Ga unit cell.

### *6.4.1 Material properties*

Microstructure of the Ni-Mn-Ga films constrained on a MgO substrate is investigated by SEM. Fig. 6.24 shows the Ni-Mn-Ga film surface on a micrometer scale. Twin boundaries are aligned by 45° along the MgO [010] and MgO [100] crystallographic direction. The twinning planes are orientated in the {011} directions.

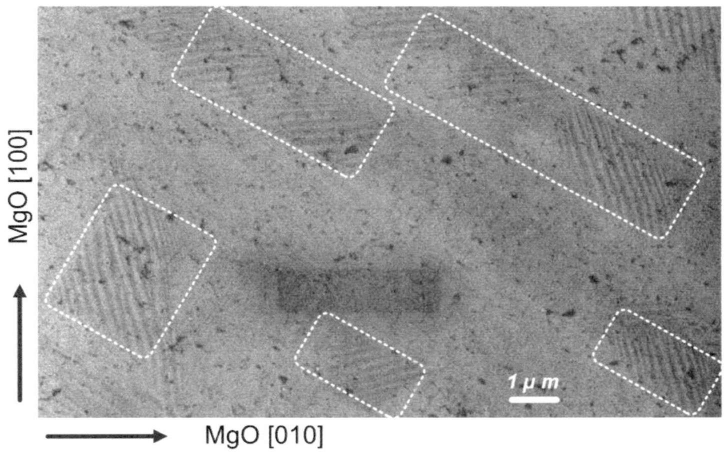


**Figure 6.24:** SEM micrograph of an epitaxial Ni-Mn-Ga film used to analyze the microstructure and twin boundary orientations of martensite variants, the localized 14M modulated twin boundaries are indicated.

On the surface of Ni-Mn-Ga film, localized regions of twin boundaries aligned by 45° are identified and are corresponding to 14M modulation as discussed in section 6.1. However, localized distinct twin morphology on the surface is identified to be quite different than sample of type 1 and 2. Furthermore, the traces of NM twin boundaries are not observed on the surface of freestanding Ni-Mn-Ga film. In order to understand the two kinds of twin microstructure corresponding to 14M and NM, one would need to study the cross-section by FIB. The film surface only shows the traces of the twin boundaries and is identified as the 14M-martensite variants periodically aligned on the surface.

The phase transformation temperatures of freestanding Ni-Mn-Ga stripes are investigated by four-probe resistance measurements. The characteristic of phase transformation between austenite and martensite is observed at high temperatures above Curie temperature of 90 °C as shown in Fig. 6.25.

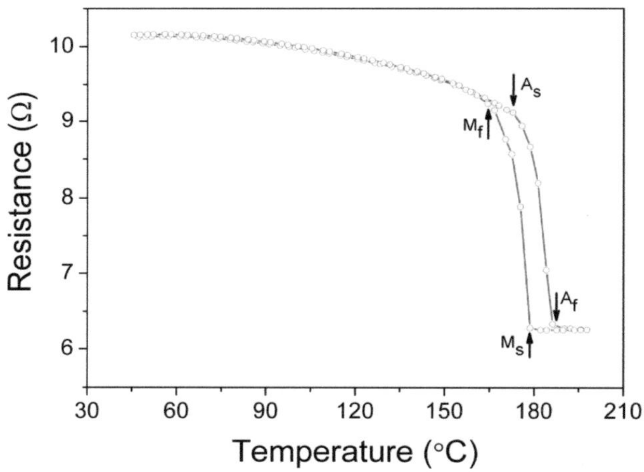


**Figure 6.25:** Electrical resistance characteristics of freestanding Ni-Mn-Ga film showing phase transformation above 150 °C.

Upon heating, the austenite start temperature at $A_s$ = 172 °C and the corresponding finish temperature at $A_f$ = 186 °C is observed. Upon cooling, martensite formation starts at $M_s$ = 178 °C and ends at $M_f$ = 165 °C, respectively. The observed high phase transformation temperatures are agreement with the NM martensite observed in bulk Ni-Mn-Ga single crystal. The characteristic high phase transformation behavior is similar to sample of type 1. Thus, by using SEM measurements the presence of 14M martensite and by resistance measurements NM martensites are confirmed. Hence the presence of both 14M and NM modulated martensites in the microstructure of samples of type 3 is similar to sample of type 1 and 2.

### *6.5.2 Mechanical properties before and after training*

Training is being performed on freestanding Ni-Mn-Ga films by applying a constant stress of 2 and 6 MPa along the tensile loading direction. Five temperature cycles are performed between martensite finish and austenite start temperatures. By using thermo-mechanical training, accommodation and reorientation of the variants takes place in the martensite regime. Engineering stress-strain curves for both untrained and trained Ni-Mn-Ga films are shown in Fig. 6.26.

Twin boundary mobility is studied by applying a tensile stress along the [001] direction of the Ni-Mn-Ga unit cell. For untrained Ni-Mn-Ga film, a high twinning stress of 28 MPa along with large superplastic strain of 14% is observed. At 250 MPa, cracking occurs in the film revealing the ultimate strength of the material. The observed high twinning stress, large superplastic strain of >11% indicate that the stress induced variant reorientation is due to coexistence of 14M and NM martensite variants.

The detwinning and inter-martensitic transition between 14M to NM martensites is explained by stress-strain measurements in section 6.2.4 for as-released Ni-Mn-Ga films. After training of Ni-Mn-Ga films, a decrease of twinning stress at low strain levels below 3 % strain is observed. It is shown in Fig. 6.26 (a) and (b). Further deformation beyond 3% strain gives rise to similar stress-strain behavior as observed before. In contrast to untrained specimen, training performed at 2 MPa yields large plastic strain up to 12 %, for 6 MPa it is about 10 %. This result indicates that training leads to formation of larger fractions of 14M martensite whose reorientation gives rise to a macroscopic change of strain. Such behavior has not been observed for samples of type 1 and 2.

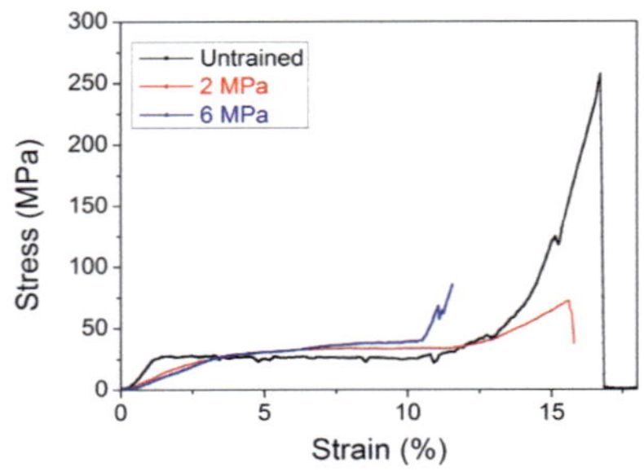


**(a)**

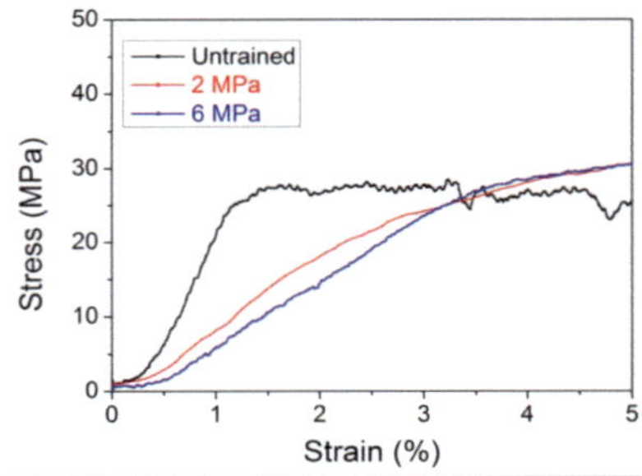


**(b)**

**Figure 6.26:** (a) Tensile stress versus strain characterstics of both untrained and trainied Ni-Mn-Ga freestanding films. (b) Magnification view of the stress-strain characterstics shown in (a).

The mechanically induced reorientation of 14M martensite twins occurs at a relatively low twining stress in the order of several MPa. Such values are already very attractive to induce magnetic field induced reorientation. Hence, such advanced training methods are effective to decrease the twinning stress in order to enable MIR. Starting from a multivariant state of a Ni-Mn-Ga film, usually one of the long axes along the tensile loading direction is aligned after thermo-mechanical training while the other axes are not affected. Hence, further training on Ni-Mn-Ga films should aim at magnetic field assisted thermo-mechanical training for better alignment of martensitic variants. So far, all attempts to induce MIR in samples of type 3 failed despite the relatively low value of twinning stress. It seems that films exhibiting mixed 14M and NM martensites are no good candidates to induce MIR. Presumably, a more promising approach is the development of Ni-Mn-Ga films with 10M modulated crystal structure combined with a suitable magnetic field assisted thermo-mechanical training method.

# Chapter 7

# Conclusions and outlook

## 7.1 Conclusions

Development of microactuators using Ni-Mn-Ga foils and films are studied. The magnetic field induced actuation in Ni-Mn-Ga foils is demonstrated to make use of the MIR effect in small-scale applications. Two different technological approaches are described. First one is to reduce thickness of FSMA bulk single crystals to prepare FSMA foils, second one is based on sputter deposition of epitaxial Ni-Mn-Ga films. In addition to this, process technologies such as sacrificial layer and sacrificial substrate technologies are developed for sputter deposited epitaxial Ni-Mn-Ga films on MgO substrate. As the film and foil actuators cannot be operated by an in-plane compressive load, it is proposed to operate the actuators by an in-plane or out-of-plane tensile load. For Ni-Mn-Ga film and foil actuators with magnetic easy axis along the short c-axis, applying the magnetic field along the tensile loading direction is desirable to induce the MIR effect.

Ni-Mn-Ga foils are prepared by electric discharge machining (ED) and abrasive wire sawing (AS) methods. Foils prepared by ED-method reveal thickness dependent microstructural and mechanical properties. XRD measurements are performed on 200 μm foils shows 10M modulated martensite similar as bulk Ni-Mn-Ga alloys; indifferently 30 μm foils reveals 14M modulated martensite. Temperature dependent phase transformation characteristics reveal a large deviation in transformation temperatures between 200-30 μm Ni-Mn-Ga foils. In as-received state, the Ni-Mn-Ga foils (200 μm thick) exhibit a relatively high initial twinning stress of 5.5 MPa. This is at least 2.5 times higher as compared to corresponding bulk Ni-Mn-Ga alloys. In addition, the twinning stress of 30 µm foil is estimated to be 23 MPa indicating severe damage to the surface of foil due to ED method. Surface twin morphology investigations reveal a distorted, deformed and random orientation of martensite twins on the surface of 30 µm foil which is not observed for 200 µm Ni-Mn-Ga foils. Therefore, the influence of surface damage on crystal structure, phase transformation as well as mechanical properties is discussed depending on the foil thickness. Different training methods are adopted to decrease the twinning stress and to improve the microstructure. In some cases, the low twinning stress of the bulk reference alloy is recovered after thermo-mechanical training in tensile mode. In this case, a partial MIR up to 1 % is observed in magneto-strain experiments at zero bias stress, but the effect of thermo-mechanical training below 200 µm thickness seems to be less significant. Further tensile experiments in combination with DIC technique is performed on 80 µm thick Ni-Mn-Ga foil. In as-received Ni-Mn-Ga foil, strain field is inhomogeneous during tensile loading and the deformation is associated with intermittent strain bands due to localized defects at the surface. After training, the foil shows almost homogeneous deformation all over the specimen surface. TEM investigations performed on 30 µm thick Ni-Mn-Ga foils probed on the surface damage layer shows clear evidence of stacking faults distributed randomly at different pinning sites. Furthermore the crystal structure at room temperature is FCC cubic with lattice constant of 3.67 Å, which is quite different compared to the bulk reference sample.

In contrast to ED method, foils prepared by AS method show similar 10M crystal structure, phase transformation and mechanical properties as of bulk Ni-Mn-Ga alloy. The linear actuation in a beam-shaped Ni-Mn-Ga foil actuator based on the MSM effect has been demonstrated. Single crystalline Ni-Mn-Ga foils of 200 µm, 30 µm are considered to study the actuation characteristics and to investigate their performance effect on thickness. XRD measurements confirm 10M modulated tetragonal martensite in a single variant state for both

200 μm and 30 μm foils. Boundary conditions for the onset stress to nucleate martensite twin variants at different magnetic field strengths have been studied in combination with tensile mechanical properties, which are crucial for optimal actuation response. A maximum strain response of ~6 % due to variant reorientation for 200 μm and 30 μm foils above a critical magnetic field of 0.12 T has been reported. However, using MSM linear actuator for practical application the maximum usable strain can be limited due to the constraints of fixation and loading. Demonstration of miniaturized linear actuator has been presented for the first time based on Ni-Mn-Ga foils, which shows repeatable full actuation cycles with a maximum actuation stroke of 2.2 % when operated at a tensile load of about 1.5 MPa. Further improvements to achieve maximum actuation stroke up to 6% are suggested, e.g., using a tensile elastic spring to reset the actuator in a magnetic field. In addition, the influence of size effects on the performance of an actuator must be understood for small-scale applications.

Critical issues involved for enabling the magnetic field induced reorientation in Ni-Mn-Ga films is discussed and investigated. Several technological approaches are developed to attain freestanding Ni-Mn-Ga films. Firstly the sacrificial substrate technology is presented to release the epitaxial Ni-Mn-Ga films deposited on NaCl substrates, but due to hydroscopic nature of the NaCl substrate quite complex surface morphology is attributed, thus this process is not quite successful. Alternatively, sacrificial layer technology is developed to release the Ni-Mn-Ga film from MgO substrate. Fabrication of freestanding epitaxial Ni-Mn-Ga films by a combination of magnetron sputtering and sacrificial layer technology is presented. The feasibility of the sacrificial layer technology process is demonstrated for an MgO (100) substrate covered by a Cr buffer layer. This new fabrication route can be applied to develop novel microactuators, such as stripe and bridge actuators. The Ni-Mn-Ga stripe actuators show the conventional shape memory effect as well as superplastic behavior due to co-existing 14M and NM martensites at room temperature.

Three different types of freestanding films are investigated distinguished by their composition, deposition temperature and thickness towards realizing the novel FSMA microactutors.

Ni-Mn-Ga samples of type 1 deposited at 400°C with $Ni_{50.5}Mn_{30.5}Ga_{19}$ composition are investigated. These films are patterned along the different crystallographic directions allowing direction-dependent tensile measurements after substrate release. X-ray diffraction results on the freestanding films show major contributions of NM and minor contributions of 14M martensite. In the as-released state, SEM-FIB cross-sections reveal two different twin patterns on a mesoscopic scale; a first twin pattern is inclined by 45° and a second twin pattern is observed parallel to the film growth direction. TEM investigations identify the twin boundaries aligned 45° with respect to the film surface as mesoscopic 14M twins. The NM structure is also reflected in electrical resistance measurements by high phase transformation temperatures. For the [100] and [110] crystallographic direction of Ni-Mn-Ga unit cell, the tensile measurements reveal superplastic behavior with a strain plateau of 12 % and 4%, respectively, indicating stress-induced reorientation of NM martensite starting from the multi-variant initial state. Due to the NM structure, a rather high twinning stress of 25 and 30 MPa is observed for the [100] and [110] directions, respectively. The orientation dependence of twinning stress can be qualitatively explained by the different orientations of external tensile load with respect to the twinning planes. The microstructure of the freestanding Ni-Mn-Ga film after large super plastic strain of 12 % along [100] direction is investigated. HRTEM and FIB-SEM results confirm that after straining up to 12%, no twin pattern is observed. The film is in a single variant state of NM martensite. Tensile loading of 25 MPa appears to be sufficient for detwinning of 14M structures, which is, according to the concept of adaptive martensite, composed of nanotwinned NM variants. The observed effect of tensile loading on phase formation has to be taken into account for the development of a training process to reduce the twinning stress and to enable magnetic field induced reorientation. Any training process should aim at supporting the formation of mesoscopic twin boundaries showing macroscopically observable high mobility. Microstructural investigation of freestanding Ni-Mn-Ga films before and after variant reorientation indicate that pure mechanical training will not be effective. Instead, combinations of thermo-mechanical or thermo-magneto-mechanical training are suggested as a route to achieve this goal.

The complex transformation behavior in samples of type 2 freestanding films is investigated. These samples are deposited at 300°C showing a different chemical composition of $Ni_{47.7}Mn_{31.1}Ga_{21.2}$, compared to samples of type 1. Temperature-dependent magnetization, electrical resistance, X-ray diffraction and stress-strain measurements are performed to understand the complex transformation between the co-existing 14M and NM martensite phases. A two-stage phase transformation in the temperature ranges (I) of 40°-80 °C, and (II) of 140°-160°C is observed. The combination of experiments allows excluding one austenite-martensite transformation. It is suggested that a first order transformation occurs in temperature regime (I) followed by an intermartensitic transformation in temperature regime (II). Though our analysis shows that the use of one single method can be misleading, even the present combination of four different temperature dependent methods is not able to prove the concept unambiguously. Further experiments are suggested using temperature dependent measurements of martensitic microstructure and XRD at different angles in order to separate changes in structure and martensitic microstructure.

Thermo-mechanical training is performed on samples of type 3, which are sputter deposited at 350°C. A different chemical composition of $Ni_{49.2}Mn_{32.4}Ga_{18.4}$ is observed compared to samples of type 1 and 2. Training seems to be partially successful on freestanding films due to coexisting 14M and NM martensite phases. A drastic decrease of twinning stress is achieved in the strain regime below 3 % strain is observed in stress-strain measurements.

However, training is not fully effective due to co-existing 14M and NM martensite state of freestanding Ni-Mn-Ga films. So far, the freestanding film actuators still do not show the right (micro-) structure to observe bulk-like mechanical behavior. For MSM actuation based on Ni-Mn-Ga films, a low twinning stress in the order of 1 MPa is required. Therefore, further improvements in the deposition techniques are necessary to fabricate pure 10M modulated films by precisely controlling the deposition parameters. Then, further investigations on such freestanding films are certainly helpful to achieve low twining stress by using the novel magneto-thermo-mechanical training methods developed in this thesis.

## 7.2 Outlook

So far no MSM microactuator is identified based on freestanding epitaxial Ni-Mn-Ga films, because of several improvements are required in preparation of modulated epitaxial Ni-Mn-Ga films, new technological approaches along with design rules. To prepare 10M modulated epitaxial Ni-Mn-Ga films with low number of defects, a deeper understanding of the influence of internal interfaces, e.g., the habit plane, on the phase formation and the effect of variant selection on the growth mechanism of Ni-Mn-Ga films are required. High resolution TEM experiments are certainly beneficial to understand structural phenomena at the nano scale regarding habit plane formation, growth mechanisms and also defect density in sputtered epitaxial Ni-Mn-Ga films.

Full or partial releasing of Ni-Mn-Ga films from the MgO substrate will be required for MSM film actuation. Three different options are available which will allow for free movement of Ni-Mn-Ga film. First, using Cr sacrificial layer technology full releasing of the Ni-Mn-Ga film from the MgO substrate is possible in order to allow for free movements of the twin boundaries. This technological approach is discussed in section 6.1. Second, full releasing of Ni-

Mn-Ga films from the substrate can also be obtained by using sacrificial substrate technology and this approach is discussed in section 4.1. Third, partial release of Ni-Mn-Ga film from the MgO substrate may be possible by using hard masking techniques. Here one can maintain stable contact of the base of Ni-Mn-Ga cantilever beam structure with the rigid MgO substrate while the movable cantilever beam is released within the substrate plane. By using hard mask as a etching barrier, partial release of the Ni-Mn-Ga film from the MgO substrate may be possible by optimising wet-chemical or dry etching processes. Overall, first and second releasing approaches needs further integration of freestanding Ni-Mn-Ga film to another carrier substrate by transfer bonding technology, which may cause blocking of twin boundaries at the bonding sites of the freestanding Ni-Mn-Ga film. In addition, integration of micro or nano MSM actuators by this approah is difficult. Hence it is suggested to use third approach such as partial release of Ni-Mn-Ga film from the MgO substrate which allows to fabricate MSM actuators at micro and nano scale with high degree of design flexibility.

MSM actuation at small samples sizes (while reducing film thickness and width significantly) have to be understood by considering size effects, which will arise when a characterstic length related to the deformation mechanics interact with a size parameter such as film thickness [124]. Regardless, the microstructure of the sample at any sample size is a key parameter for MSM actuation. Large change of mechanical properties is expected when the sample size is comparable to the characteristic length relevant for the primary deformation mechanism. Research on bulk single crystalline MSM material indicates that variant reorientation is a dislocation-mediated process involving characterstic lengths of 1 μm [125]. It would be interesting to investigate such size effects for MSM epitaxial thin films and related nanostructured MSM film devices.

Ganor et al [105] demonstrated an increase in the blocking stress by more than 100% by reducing the Ni-Mn-Ga actuator size by square cross-section to 200 μm. This result is interpreted by an increased value of the blocking stress due to enhancement in energy barrier to magnetization rotation. Therefore, the work output of MSM actuator is expected to significantly increase at small scales. This indicates a fundamental relationship among the specimen size, its microstructure, and its physical properties. Heczko et al [16] suggested that twin boundary mobility depends on Ni-Mn-Ga sample volume and that the switching field for MSM actuation drastically decreases for decreasing sample size by its thickness. Because the mobility of twin boundaries is hindered by pinning sites at internal obstacles, the decrease in volume will conversely reduce the pinning sites, and thus the MSM effect can occur at lower magnetic field.

Another important size effect in Ni-Mn-Ga films is due to the influence of magnetic stray field energy on variant reorientation in a magnetic field. In Ni-Mn-Ga films, variants having their easy magnetization axis perpendicular to the film plane are energetically disfavoured, since they result in a high magnetic stray field energy. This is not the case for variants with their easy axis in-plane, where the formation of magnetic domains allows flux closure and reduces magnetostatic energy. Thus, at small scales once the single magnetic domain state is approached the relative stray field energy contribution increases [126]. This increased energy input for actuation provides a scaling behavior that is beneficial for sub-micrometer devices [14]. Since for sub-micron devices a single domain martensite nuclei is possible which may reduce their stray field energy significantly during the martensitic transformation by aligning their easy axis in-plane to acheive flux closure with the neighboring austenite. This stray field induced martensite (SFIM) benefits from the combination of ferromagnetic and martensitic properties present in MSM materials. Since a reduced stray field energy induces the uneven variant distribution, no mechanical training is required as for conventional two way shape memory alloys.

Once a low twinning stress of below 5 MPa is achieved in freestanding epitaxial Ni-Mn-Ga films several novel ways may be further investigated to probe MIR. One possibility is the crystallographic analysis of Ni-Mn-Ga films with and without an applied magnetic field by x-ray diffraction. The reorientation of martensite variants in a magentic field could be identified both in-situ and ex-situ by comparing the intensities and orientation distribution of the presence of crystallographic phases in Ni-Mn-Ga films. By using Digital Image Correlation (DIC) the local strain distributions with sub-micron scale resolutions can be tracked in a Ni-Mn-Ga freestanding film. DIC can be performed in-situ to probe MIR in freestanding Ni-Mn-Ga film in an external magnetic field.

After realizing magnetic field induced reorientation in Ni-Mn-Ga freestanding films, there are several options for the development of applications that use the advantage of the MSM effect in small dimensions. MSM micro and nanoactuators have unique advantages due two main reasons.

1. Upon miniaturization, the mechanical properties of most competing actuation principles show a minor down scaling behavior. For MSM actuators, the force scales proportional to $r^2$ (r is size of an actuator). Therefore, the work density is expected to scale independent of the size of the actuators, when using an external magnetic field source and neglecting material-dependent size effects.
2. The small strain of piezoelectric actuators, e.g., of PZT in the order of 0.1 % can not be enlarged by gear mechanisms in small dimensions due to limited space. Consequently, MSM actuators become increasingly attractive for application upon miniaturization.

One possible of such application in telecommunication could be optical switching based on MSM effect. A possible layout is illustrated by the schematic shown in Fig. 7.1.

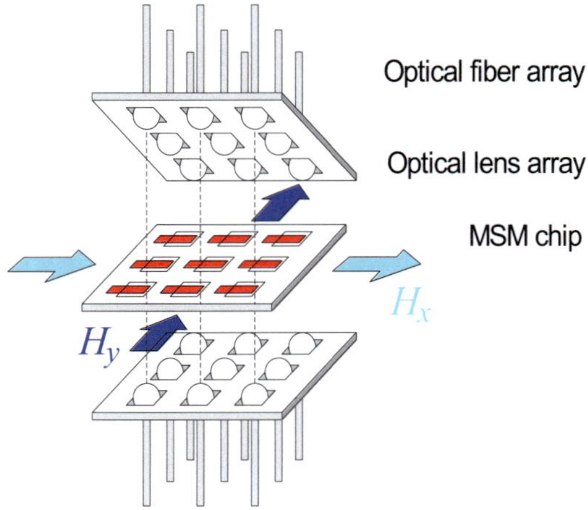


**Figure 7.1:** Integrated MSM optical microchip. The optical beam path is illustrated for the first row by dashed lines. $H_x$ and $H_y$ denote counteracting in-plane magnetic fields generated by an external field source.

The optical switch consists of a MSM chip, optical lens array and an optical fiber array. The FSMA chip consists of a series of microactuators and each actuator controls the size of an optical slit depending on the direction of the external magnetic field. The magnetic field is generated by an external magnetic field source in both counteracting magnetic fields along $H_x$ and $H_y$ directions as shown in Fig. 7.1. Since the present application involves optical switching using MSM linear actuation is considered, so the present example does not require any mechanical loading or electrical contacts. Hence this application fulfills the major criteria for further miniaturization. MSM effect can introduce a large linear displacement for a given size of an individual microactuator, which is considered unique for this kind of matrix switch. As the magnetic field is not integrated in the system, the MSM chip may be easily replaced or exchanged for different switching applications.

Depending on the technical application, other actuation functions can be implemented. One option is to implement an application of external mechanical load to each MSM microactuator as discussed in chapter 4. This approach would open up the possibility to tune the stress state of the MSM film actuator relatively to the blocking stress as well as to realize integrated training on the chip. Further miniaturization will eventually lead to novel nanoactautor arrays, e.g., for nanoplasmonic applications. For example metamaterials use an electromagnetic resonator in order to obtain a negative diffraction index. Integrating an FSMA nanoactuator within the resonance circuit could allow adjusting the frequency in a remote way in a reversible manner. An external magnetic field could be used to adjust the direction of the easy axis in all such kinds of application. Other potential applications would be tunable nanoantennas, plasmonic tweezers or spacers.

# References

[1] H. Janocha (Ed.), *Actuators: Basics and Applications*, Springer, ISBN 3-540-61564-4 (2004).

[2] H. Janocha, *Adaptronics and Smart Structures: Basics, Materials, Design, and Applications*, Springer, ISBN 3-540-61484-2 (1999).

[3] M. Kohl, *Shape Memory Microactuators*, Springer Verlag Berlin Heidelberg, (2004).

[4] E. Thielicke, E. Obermeier, *Microactuators and their technologies*, Mechatronics, 10 (2000) 431-455.

[5] G. Engdahl, *Handbook of giant magnetostrictive materials*, Academic Press, Sandiego, (2000) p.386.

[6] A. G. Olabi, A. Grunwald, *Design and applycations of magnetostrictive materials* Materials & Design, 29 (2008) 469-483.

[7] K. Ullakko, J.K. Huang, C. Kantner, R.C. O'Handley and V.V. Kokorin, *Large magnetic field-induced strains in Ni2MnGa single crystals*, Applied Physics Letters, 13 (1996) 1966-1968.

[8] R.D. James, M. Wuttig, *Magnetostriction of martensite*, Philosophical Magazine A, 77 (1998) 1273-1299.

[9] E. Quandt, M. Wuttig, *Comparision of joule and twin induced magnetostriction*, Proc. of Actuator 2004 Conference, (2004) pp. 359-362.

[10] V.A. Chernenko, V.L'Vov, J. Pons and E. Cesari, *Supereleasticity in high-temperature Ni-Mn-Ga alloys*, Journal of Applied Physics, 93 (2003) 2394-2399.

[11] I. Suorsa, J. Tellinen, K. Ullakko, E. Pagounis, *Voltage generation induced by mechanical straining in magnetic shape memory materials*, Journal of Applied Physics, 95 (2004) 8054-8058.

[12] M. Kohl, Y.S. Reddy, F. Khelfaoui, B. Krevet, A. Backen, S. Fahler, T. Eichhorn, G. Jakob, A. Mecklenburg, *Recent Progress in FSMA Microactuator Developments*, Ferromagnetic Shape Memory Alloys, 635 (2010) 145-154.

[13] M. Kohl, E. Just, W. Pfleging, S. Miyazaki, *SMA microgripper with integrated antagonism*, Sensor and Actuators, 83 (2000) 208-213.

[14] M. Thomas, O. Heczko, J. Buschbeck, Y.W. Lai, J. McCord, S. Kaufmann, L. Schultz, S. Fahler, *Stray-Field-Induced Actuation of Free-Standing Magnetic Shape-Memory Films*, Advanced Materials, 21 (2009) 3708.

[15] R. Niemann, O. Heczko, L. Schultz, and S. Fähler, *Metamagnetic transitions and magnetocaloric effect in epitaxial Ni-Co-Mn-In films*, Applied Physics Letters, 97 (2010) 222507.

[16] O. Heczko, A. Soroka, S.-P. Hannula, *Magnetic shape memory effect in thin foils*, Applied Physics Letters, 93 (2008).

[17] R.C. O'Handley, *Model for strain and magnetization in magnetic shape-memory alloys*, Journal of Applied Physics, 83 (1998) 3263.

[18] K. Otsuka, C. M. Wayman, *Shape Memory Materials*, Cambridge University Press, 1st Edition (1998)

[19] W. J. Buehler, J. V. Gilfrich, and R.C Wiley, *Effect of low-temperature phase changes on the mechanical properties of allyos near composition TiNi*, Journal of Applied Physics, 34 (1963) 1475-1477.

[20] P. Krulevitch, A.P. Lee, P.B. Ramsey, J.C. Trevino, J. Hamilton, and M.A Northrup, *Thin film shape memory alloys microactuators*, Journal of Microelectromechanical Systems, 5 (1996) 270-282.

[21] K. Ullakko, *Magnetically controlled shape memory alloys: A new class of actuator materials*, Journal of Materials Engineering and Performance, 5 (1996) 405-409.

[22] S.J. Murray, M. Marioni, P.G. Tello, S.M. Allen, R.C. O'Handley, *Giant magnetic-field-induced strain in Ni-Mn-Ga crystals: Experimental results and modeling*, Journal of Magnetism and Magnetic Materials, 226 (2001) 945-947.

[23] S.J. Murray, M. Marioni, S.M. Allen, R.C. O'Handley, T.A. Lograsso, *6% magnetic-field-induced strain by twin-boundary motion in ferromagnetic Ni-Mn-Ga*, Applied Physics Letters, 77 (2000) 886-888.

[24] A. Sozinov, A.A. Likhachev, N. Lanska, K. Ullakko, *Giant magnetic-field-induced strain in NiMnGa seven-layered martensitic phase*, Applied Physics Letters, 80 (2002) 1746-1748.

[25] P. Müllner, V.A. Chernenko, and G. Kostorz, *Large cyclic magnetic field induced deformation in orthorhombic (14M) Ni-Mn-Ga martensite*, Journal of Applied Physics, 95 (2004) 1531.

[26] K. Koho, O. Soderberg, N. Lanska, Y. Ge, X. Liu, L. Straka, J. Vimpari, O. Heczko, V.K. Lindroos, *Effect of the chemical composition to martensitic transformation in Ni-Mn-Ga-Fe alloys*, Materials Science and Engineering A-Structural Materials Properties Microstructure and Processing, 378 (2004) 384-388.

[27] V. Sanchez-Alarcos, V. Recarte, J.I. Perez-Landazabal, C. Gomez-Polo, V.A. Chernenko, M.A. Gonzalez, *Reversible and irreversible martensitic transformations in Fe-Pd and Fe-Pd-Co alloys*, European Physical Journal-Special Topics, 158 (2008) 107-112.

[28] A.N. Lavrov, S. Komiya, and Y. Ando, *Antiferromagnets: Magnetic shape-memory effects in a crystal*, Nature, 418 (202) 385-386.

[29] S.J. Murray, R. Hayashi, M. Marioni, S.M. Allen, R.C. O'Handley, *Magnetic and mechanical properties of FeNiCoTi and NiMnGa magnetic shape memory alloys*, Smart Structures and Materials 1999: Smart Materials Technologies, 3675 (1999) 204-211.

[30] R. Hayashi, S.J. Murray, M. Marioni, S.M. Allen, R.C. O'Handley, *Magnetic and mechanical properties of FeNiCoTi magnetic shape memory alloy*, Sensors and Actuators A -Physical, 81 (2000) 219-223.

[31] T. Kakeshita, K. Ullakko, *Giant magnetostriction in ferromagnetic shape-memory alloys*, MRS Bulletin, 27 (2002) 105-109.

[32] N. Singh, E. Dogan, I. Karaman, R. Arroyave, *Effect of configural order on the magnetic characterstics of Co-Ni-Ga ferromagnetic shape memory alloys*, Phys. Rev. B, 84 (2011) 184201.

[33] M. Acet, E. Duman, E. F. Wassermann, L. Manosa, A. Planes, *Coexisting ferro- and antiferromagnetism in Ni2MnAl Heusler alloys*, Journal of Applied Physics, 92 (2002) 3867.

[34] Y. Murakami, D. Shindo, K. Oikawa, R. Kainuma, K. Ishida,, *Magnetic domain structures in Co-Ni-Al shape memory alloys studied by Lorentz microscopy and electron holography*, Acta Materialia, 50 (2002) 2173-2184.

[35] D.C. Dunand, P. Müllner, *Size Effects on Magnetic Actuation in Ni-Mn-Ga Shape-Memory Alloys*, Advanced Materials, 23 (2011) 216-232.

[36] I. Suorsa, E. Pagounis, K. Ullakko, *Magnetization dependence on strain in the Ni-Mn-Ga magnetic shape memory material*, Applied Physics Letters, 84 (2004) 4658-4660.

[37] A. Sozinov, P. Yakovenko, K. Ullakko, *Large magnetic-field-induced strains in Ni-Mn-Ga alloys due to redistribution of martensite variants*, European Magnetic Materials and Applications, 373-3 (2001) 35-39.

[38] L. Straka, O. Heczko, K. Ullakko, *Investigation of magnetic anisotropy of Ni-Mn-Ga seven-layered orthorhombic martensite*, Journal of Magnetism and Magnetic Materials, 272 (2004) 2049-2050.

[39] S.J. Murray, M. Farinelli, C. Kantner, J.K. Huang, S.M. Allen, R.C. O'Handley, *Field-induced strain under load in Ni-Mn-Ga magnetic shape memory materials*, Journal of Applied Physics, 83 (1998) 7297-7299.

[40] Y. Ma, S. Awaji, K. Watanabe, M. Matsumoto, N. Kobayashi, *X-ray diffraction study of the structural phase transition of Ni2MnGa alloys in high magnetic fields*, Solid State Communications, 113 (2000) 671.

[41] P.J. Brown, J. Crangle, T. Kanomata, M. Matsumoto, K.U. Neumann, B. Ouladdiaf, K.R.A. Ziebeck, *The crystal structre and phase transitions of the*

*magnetic shape memory compounds Ni2MnGa*, J. Phys.: Condens. Mater, 14 (2002) 10159.

[42] G.B. Olson, W.S. Owen (Editor), *Martensite*, ASM International, (1992).

[43] O. Heczko, L. Straka, N. Lanska, K. Ullakko, J. Enkovaara, *Temperature dependence of magnetic anisotropy in Ni-Mn-Ga alloys exhibiting giant field-induced strain*, Journal of Applied Physics, 91 (2002) 8228-8230.

[44] V.V. Martynov, and V.V. Kokorin, *The crystal structure of thermally and stress induced martensites in Ni-Mn-Ga single crystals*, Journal De Physique Iv, 2 (1992) 739-749.

[45] V.A. Chernenko, E. Cesari, V.V. Kokorin, I.N. Vitenko, *The development of new ferromagnetic shape memory alloys in Ni-Mn-Ga system*, Scripta Metallurgica Et Materialia, 33 (1995) 1239-1244.

[46] V. Soolshenko, N. Lanska, K. Ullakko, *Structure and twinning stress of martensites in non-stoichiometric Ni2MnGa single crystal*, Journal De Physique Iv, 112 (2003) 947-950.

[47] V.A. Chernenko, J. Pons, C. Segui and E. Cesari, *Premartensitic phenomena and other phase transformations in Ni-Mn-Ga alloys studied by dynamical mechanical analysis and electron diffraction*, Acta Materialia, 50 (2002) 53-60.

[48] V.A. Chernenko, J. Pons and E. Cesari, *Premartensitic phenomena and other phase transformations in Ni-Mn-Ga alloys studied by dynamical analysis and electron diffraction*, Acta Materialia, 50 (2002) 53-60.

[49] O. Heczko, N. Lanska, O. Soderberg, K. Ullakko, *Temperature variation of structure and magnetic properties of Ni-Mn-Ga magnetic shape memory alloys*, Journal of Magnetism and Magnetic Materials, 242 (2002) 1446-1449.

[50] V.A. Chernenko, E. Cesari, V.V. Kokorin, I.N. Vitenko, *The Development of new ferromagnetic shape-memory allyos in Ni-Mn-Ga system,* Scripta Metallurgica Et Materialia, 33 (1995) 1239-1244.

[51] J. Pons, V.A. Chernenko, R. Santamarta and E. Cesari, *Crystal structure of martensitic phases in Ni-Mn-Ga shape memory alloys*, Acta Materialia, 48 (2000) 3027-3028.

[52] R.C. O'Handley, and S.M. Allen, *Shape memory alloys, magnetically activated ferromagnetic shape-memory materials*, In: M. Schwartz (Ed.) Encyclopedia of smart Materials, John Wiley and Sons, New York, (2001) pp. 936-951.

[53] K. Tsuchiya, H. nakamura, D. Ohtoyo, H. nakayama, H. Ohtsuka and M. Umemoto, *Composition dependence of phase transformations and microstructures in Ni-Mn-Ga ferromagnetic shape memory alloys*, Proceedings of Internatinal Symposium on Designing, Processing and Properties of Advanced Engineering Materials (ISAEM), (2000) pp. 409-414.

[54] P.J. Webster, K.R.A. Ziebeck, S.L. Town, M.S. Peak, *Magnetic order and phase transformation in Ni2MnGa*, Philosophical Magazine B, 49 (1984) 295.

[55] N. Glavatska, G. Mogylny, I. Glavatsky, A. Tyshchenko, O. Söderberg and V.K. Lindroos, *Temperature dependence of magnetic shape memory effect and martensitic structure of NiMnGa alloy*, Material Science Forum, 394-395 (2002) 537-540.

[56] O. Heczko, A. Sozinov, K. Ullakko, *Giant field-induced reversible strain in magnetic shape memory NiMnGa alloy*, IEEE Transactions on Magnetics, 36 (2000) 3266-3268.

[57] A. Sozinov, A.A. Likhachev, N. Lanska, K. Ullakko, V.K. Lindroos, *Crystal structure, magnetic anisotropy and mechanical properties of seven-layered martensite in Ni-Mn-Ga*, Smart Structures and Materials 2002: Active Materials: Behavior and Mechanics, 4699 (2002) 195-205.

[58] G. Fritsch, V.V. Kokorin, V.A. Chernenko, A. Kempf and I.K. Zasimchuk, *Martensitic transformation in Ni-Mn-Ga allyos*, Phase Transitions, 57 (1996) 233-240.

[59] V.A. Chernenko, V. L'vov, E. Cesari, J. Pons, R. Portier and S.

Zagorodnyuk, *New aspects of structural and magnetic behavior of martensites in Ni-Mn-Ga alloys*, Materials Transactions, 43 (2002) 2394-2399.

[60] X.W. Liu, O. Soderberg, Y. Ge, N. Lanska, K. Ullakko, V.K. Lindroos, *On the corrosion of non-stoichiometric martensitic Ni-Mn-Ga alloys*, Journal De Physique IV, 112 (2003) 935-938.

[61] K. Suorsa, E. Pagounis, K. Ullakko, *Magnetic shape memory actuator performance*, Journal of Magnetism and Magnetic Materials, 272 (2004) 2029-2030.

[62] J. Tellinen, I. Suorsa, E. Pagounis, I. Aaltio, and K. Ullakko, *Applications of magnetic shape memory actuators*, Proc. of Actuator 2002 Conference, p. 158.

[63] B. Winzek, S. Schmitz, H. Rumpf, T. Sterzl, R. Hassdorf, S. Thienhaus, J. Feydt, M. Moske, and E. Quandt, *Recent developments in shape-memory thin film technology*, Material Science and Engineering A, 378 (2004) 40-46.

[64] X.Y. Dong, C. Adelmann, J. Palmstrom, X. Lou, J. Strand, P.A. Crowell, J.P. Barnes, and A.K. Petford-Long, *Spin injection from the Heusler alloy Co2MnGe into Al0.1Ga0.9As/GaAs heterostructures*, Applied Physics Letters, 86 (2005) 102107.

[65] P.J. Webster, *Magnetic and chemical order in Heusler alloys containing cobalt and manganese*, Journal of Physics and Chemestry of Solids, 32 (1971) 1221-1231.

[66] J.M. Guldbakke, K. Rolfs, A. Mecklenburg, J. Hesselbach, A. Raatz, R. Schneider, *A Miniaturized Gripper Driven by a Magnetic Shape Memory Single Crystal Actuator*, Proceedings of Actuator 2008, (2008) 880.

[67] M. Kohl, B. Krevet, M. Ohtsuka, D. Brugger, Y. Liu, *Ferromagnetic shape memory microactuators*, Materials Transactions, 47 (2006) 639-644.

[68] J.W. Dong, J. Lu, J.Q. Xie, L.C. Chen, R.D. James, S. McKernan, C.J. Palmstrom, *MBE growth of ferromagnetic single crystal Heusler alloys on (001)Ga1-xInxAs*, Physica E, 10 (2001) 428-432.

[69] A. Hakola, O. Heczko, A. Jaakola, T. Kajava, K. Ullakko, *Ni-Mn-Ga films on Si, GaAs and Ni-Mn-Ga single crystals by pulsed laser deposition*, Applied Surface Science, 238 (2004) 155-158.

[70] Y. Zhang, R.A. Hughes, J.F. Britten, W. Gong, J.S. Preston, G.A. Botton, M. Niewczas, *Epitaxial Ni-Mn-Ga films derived through high temperature in situ depositions*, Smart Materials & Structures, 18 (2009).

[71] M. Suzuki, M. Ohtsuka, M. Matsumoto, Y. Murakami, D. Shindo, K. Itagaki, *Effect of aging time on shape memory properties of sputtered Ni-rich Ni2MnGa alloy films*, Materials Transactions, 43 (2002) 861-866.

[72] S. Besseghini, A. Gambardella, V.A. Chernenko, M. Hagler, C. Pohl, P. Müllner, M. Ohtsuka, S. Doyle, *Transformation behavior of Ni-Mn-Ga/Si(100) thin film composites with different film thicknesses*, The European Physical Journal Special Topics, 158 (2008) 179-185.

[73] J.W. Dong, L.C. Chen, C.J. Palmstrom, R.D. James, S. McKernan, *Molecular beam epitaxy growth of ferromagnetic single crystal (001) Ni2MnGa on (001) GaAs*, Applied Physics Letters, 75 (1999) 1443-1445.

[74] J.W. Dong, L.C. Chen, S. McKernan, J.Q. Xie, M.T. Figus, R.D. James, C.J. Palmstrom, *Formation and characterization of single crystal Ni2MnGa thin films*, Materials of Smart Systems III, 604 (2000) 297-302.

[75] L.C. Chen, J.W. Dong, B.D. Schultz, C.J. Palmstrom, J. Berezovsky, A. Isakovic, P.A. Crowell, N. Tabat, *Epitaxial ferromagnetic metal/GaAs(100) heterostructures*, Journal of Vacuum Science & Technology B, 18 (2000) 2057-2062.

[76] J.W. Dong, L.C. Chen, J.Q. Xie, T.A.R. Muller, D.M. Carr, C.J. Palmstrom, S. McKernan, Q. Pan, R.D. James, *Epitaxial growth of ferromagnetic Ni2MnGa on GaAs(001) using NiGa interlayers*, Journal of Applied Physics, 88 (2000) 7357-7359.

[77] M. S. Lund, J.W. Dong, J. Lu, X.Y. Dong, C. J. Palmstrom, and C. Leighton, *Anomalous magnetotransport properties of epitaxial full Heusler alloys*, Applied Physics Letters, 80 (2002) 4798-4800.

[78] O. Heczko, M. Thomas, J. Buschbeck, L. Schultz, S. Fähler, *Epitaxial Ni–Mn–Ga films deposited on $SrTiO_3$ and evidence of magnetically induced reorientation of martensitic variants at room temperature*, Applied Physics Letters, 92 (2008) 072502.

[79] F. Khelfaoui, M. Kohl, J. Buschbeck, O. Heczko, S. Fahler, L. Schultz, *A fabrication technology for epitaxial Ni-Mn-Ga microactuators*, European Physical Journal-Special Topics, 158 (2008) 167-172.

[80] M. Thomas, O. Heczko, J. Buschbeck, U.K. Rößler, J. McCord, N. Scheerbaum, L. Schultz, S. Fähler, *Magnetically induced reorientation of martensite variants in constrained epitaxial Ni–Mn–Ga films grown on MgO(001)*, New Journal of Physics, 10 (2008) 023040.

[81] M. Thomas, O. Heczko, J. Buschbeck, L. Schultz, S. Fähler, *Stress induced martensite in epitaxial Ni–Mn–Ga films deposited on MgO(001)*, Applied Physics Letters, 92 (2008) 192515.

[82] S. Doyle, V.A. Chernenko, S. Besseghini, A. Gambardella, M. Kohl, P. Muellner, M. Ohtsuka, *Residual stress in Ni-Mn-Ga thin films deposited on different substrates*, European Physical Journal-Special Topics, 158 (2008) 99-105.

[83] V.A. Chernenko, V. Golub, J.M. Barandiaran, O.Y. Salyuk, F. Albertini, L. Righi, S. Fabbrici, M. Ohtsuka, *Magnetic anisotropies in Ni-Mn-Ga films on MgO(001) substrates*, Applied Physics Letters, 96 (2010).

[84] S. Kaufmann, U.K. Rößler, O. Heczko, M. Wuttig, J. Buschbeck, L. Schultz, S. Fähler, *Adaptive Modulations of Martensites*, Physical Review Letters, 104 (2010).

[85] S. Kaufmann, R. Niemann, T. Thersleff, U.K. Rößler, O. Heczko, J. Buschbeck, B. Holzapfel, L. Schultz, S. Fähler, *Modulated martensite: why it forms and why it deforms easily*, New Journal of Physics, 13 (2011) 053029.

[86] O. Soderberg, A. Sozinov, N. Lanska, Y. Ge, V.K. Lindroos, S.P. Hannula, *Effect of intermartensitic reaction on the co-occurrence of the magnetic and structural transition in Ni-Mn-Ga alloys*, Materials Science and Engineering A-Structural Materials Properties Microstructure and Processing, 438 (2006) 957-960.

[87] R.F. Hamilton, S. Dilibal, H. Sehitoglu, H.J. Maier, *Underlying mechanism of dual hysteresis in NiMnGa single crystals*, Materials Science and Engineering A-Structural Materials Properties Microstructure and Processing, 528 (2011) 1877-1881.

[88] T. Ohba, N. Miyamoto, K. Fukuda, T. Fukuda, T. Kakeshita and K. Kato, *Fundamental structure of a Ni2MnGa intermediate phase having an orthombic lattice*, Smart Materials & Structures, 14 (2005) S197.

[89] K. Rolfs, A. Mecklenburg, J.M. Guldbakke, R.C. Wimpory, A. Raatz, J. Hesselbach, R. Schneider, *Crystal quality boosts responsiveness of magnetic shape memory single crystals*, Journal of Magnetism and Magnetic Materials, 321 (2009) 1063-1067.

[90] Christoph Eberl, Daniel Gianola, *Free DIC/DDIT functions available for MATLAB*,http://www.mathworks.com/matlabcentral/fileexchange/12413,(2010).

[91] Daniel Gianola, *Deformation mechanisms in nanocrystalline aluminum thin films: An experimental investigation*, PhD thesis, John Hopkins University (2007).

[92] M. Chmielus, C. Witherspoon, K. Ullakko, P. Muellner, R. Schneider, *Effects of surface damage on twinning stress and the stability of twin microstructures of magnetic shape memory alloys*, Acta Materialia, 59 (2011) 2948-2956.

[93] M. Chmielus, I. Glavatskyy, J.U. Hoffmann, V.A. Chernenko, R. Schneider, P. Muellner, *Influence of constraints and twinning stress on magnetic field-induced strain of magnetic shape-memory alloys*, Scripta Materialia, 64 (2011) 888-891.

[94] M. Chmielus, K. Rolfs, R. Wimpory, W. Reimers, P. Muellner, R. Schneider, *Effects of surface roughness and training on the twinning stress of Ni-Mn-Ga single crystals*, Acta Materialia, 58 (2010) 3952-3962.

[95] F. Khelfaoui, Y.S. Reddy, M. Kohl, A. Mecklenburg, R. Schneider, *Magnetic-Field-Induced Reorientation in Single Crystalline Ni-Mn-Ga Foil Actuators*, Esomat 2009 - 8th European Symposium on Martensitic Transformations, (2009).

[96] P. Müllner, A.H. King, *Deformation of hierarchically twinned martensite*, Acta Materialia, 58 (2010) 5242-5261.

[97] G. Murasawa, S. Yoneyama, T. Sakuma, *Nucleation, bifurcation and propagation of local deformation arising in NiTi shape memory alloy*, Smart Materials and Structures, 16 (2007) 160-167.

[98] L. Straka, N. Lanska, K. Ullakko, A. Sozinov, *Twin microstructure dependent mechanical response in Ni-Mn-Ga single crystals*, Applied Physics Letters, 96 (2010).

[99] L. Straka, H. Hanninen, O. Heczko, *Temperature dependence of single twin boundary motion in Ni-Mn-Ga martensite*, Applied Physics Letters, 98 (2011).

[100] Y. Ge, H. Jiang, A. Sozinov, O. Soderberg, N. Lanska, J. Keranen, E.I. Kauppinen, V.K. Lindroos, S.P. Hannula, *Crystal structure and macrotwin interface of five-layered martensite in Ni-Mn-Ga magnetic shape memory alloy*, Materials Science and Engineering a-Structural Materials Properties Microstructure and Processing, 438 (2006) 961-964.

[101] L. Straka, H. Hanninen, N. Lanska, A. Sozinov, *Twin interaction and large magnetoelasticity in Ni-Mn-Ga single crystals*, Journal of Applied Physics, 109 (2011).

[102] S.J. Murray, S.M. Allen, R.C. O'Handley, *Modeling and experiments for deformation under load in Ni-Mn-Ga ferromagnetic shape memory alloy*, Materials of Smart Systems III, 604 (2000) 279-284.

[103] S.J. Murray, M.A. Marioni, A.M. Kukla, J. Robinson, R.C. O'Handley, S.M. Allen, *Large field induced strain in single crystalline Ni-Mn-Ga ferromagnetic shape memory alloy*, Journal of Applied Physics, 87 (2000) 5774-5776.

[104] B. Krevet, V. Pinneker and M. Kohl, *A Magnetic shape memory foil actuator loaded by a spring*, Smart Materials & Structures, 21 (2012) 094013.

[105] Y. Ganor, D. Shilo, T. W. Shield and R.D. James, *Breaching the work output limitation of ferromagnetic shape memory alloys*, Applied Physics Letters, 93 (2008) 122509.

[106] T. Eichhorn, G. Jakob, *Structural and Magnetic Properties of Epitaxial Ni2MnGa Thin Films*, Materials Science Forum, 635 (2009) 155-160.

[107] S.R. Yeduru, A. Backen, S. Fahler, A. Diestel, L. Schultz, M. Kohl, *Sacrificial Layer Technology - A Route Towards Freestanding Ni-Mn-Ga Films*, Actuator 10, Conference Proceedings, (2010) 765-768.

[108] S. Fahler, O. Heczko, M. Thomas, R. Niemann, J. Buschbeck, L. Schultz, *Recent Progress Towards Active Epitaxial Ni-Mn-Ga Magnetic Shape Memory Films*, Actuator 08, Conference Proceedings, (2008) 754-757.

[109] T. Grund, R. Guerre, M. Despont, M. Kohl, *Transfer bonding technology for batch fabrication of SMA microactuators*, The European Physical Journal Special Topics, 158 (2008) 237-242.

[110] A. Backen, S.R. Yeduru, M. Kohl, S. Baunack, A. Diestel, B. Holzapfel, L. Schultz, S. Fahler, *Comparing properties of substrate-constrained and freestanding epitaxial Ni-Mn-Ga films*, Acta Materialia, 58 (2010) 3415-3421.

[111] V.A. Chernenko, M. Chmielus, P. Muellner, *Large magnetic-field-induced strains in Ni-Mn-Ga nonmodulated martensite (vol 95, 104103, 2008)*, Applied Physics Letters, 95 (2009).

[112] S.J. Murray, S.M. Allen, R.C. O'Handley, T.A. Lograsso, *Magneto-mechanical performance and mechanical properties of Ni-Mn-Ga ferromagnetic shape memory alloys*, Smart Structures and Materials 2000 - Active Materials: Behavior and Mechanics, 3992 (2000) 387-395.

[113] V.A. Chernenko, M. Chmielus, P. Muellner, *Large magnetic-field-induced strains in Ni-Mn-Ga nonmodulated martensite (vol 95, 104103, 2009)*, Applied Physics Letters, 95 (2009).

[114] V.A. Chernenko, *Compositional instability of beta-phase in Ni-Mn-Ga alloys*, Scripta Materialia, 40 (1999) 523-527.

[115] N. Lanska, O. Soderberg, A. Sozinov, Y. Ge, K. Ullakko, V.K. Lindroos, *Composition and temperature dependence of the crystal structure of Ni-Mn-Ga alloys*, Journal of Applied Physics, 95 (2004) 8074-8078.

[116] G. Gottstein, *Physikalische Grundlagen der Materialkunde*, Springer, 3rd Edition (2007).

[117] G. Mahnke, M. Seibt, S. Mayr, *Microstructure and twinning in epitaxial NiMnGa films*, Physical Review B, 78 (2008).

[118] S. Kaufmann-Weiss, M.E. Gruner, A. Backen, L. Schultz, P. Entel, S. Fähler, Physical Review Letters, 107 (2011) 206105.

[119] S.R. Yeduru, A. Backen, C. Kübel, D. Wang, T. Scherer, S. Fähler, L. Schultz and M .Kohl, *Microstructure of free-standing epitaxial Ni-Mn-Ga films before and after variant reorientation*, Scripta Materialia, 66 (2012) 566-569.

[120] Khachaturyan A G, Shapiro S M and Semenovskaya S, Physical Review B, 43 (1991) 10832.

[121] K. Bhattacharya, Microstructure of martensite: why it forms and how it gives rise to the shape memory effect? Press (2003) Chapter 10.

[122] A. Ustinov, L. Olikhovska, N. Glavaska and I. Glavatskyy, Journal of Applied Crystallography, 42 (2009) 211.

[123] K. Ullakko, Y. Ezer, A. Sozinov, G. Kimmel, P. Yakovenko, and V.K. Lindroos, *Magnetic-field-induced strains inpolycrystalline Ni-Mn-Ga at room temperature*, Scripta Materialia, 44 (2001) 475-480.

[124] D.C. Dunand, P. Mullner, *Size Effects on Magnetic Actuation in Ni-Mn-Ga Shape-Memory Alloys*, Advanced Materials, 23 (2011) 216-232.

[125] E. Arzt, *Size effects in material due to microstructural and diemensional constraints: a comparative review*, Acta Materialia, 46 (1998) 5611.

[126] A. Hubert, R. Schäfer, *Magnetic Domains*, Springer, (1998) 306.

# Publication list

***Journals:***

- A. Backen, **S.R. Yeduru,** M. Kohl, S. Baunack, A. Diestel, B. Holzapfel, L. Schultz, S. Fähler "Comparing the properties of substrate-constrained and freestanding epitaxial Ni-Mn-Ga films" *Acta Materialia 58* (2010) 3415-3421.

- M. Kohl, **S.R. Yeduru**, F. Khelfaoui, B. Krevet, A. Backen, S. Fähler, T. Eichhorn, G. Jakob and A. Mecklenburg, "Recent Progress in FSMA Microactuator Developments" *Materials Science Forum* 635 (2010) 145-154.

- **S.R. Yeduru**, A. Backen, S. Fähler, L. Schultz and M. Kohl "Large superplastic strain in Non-Modulated epitaxial Ni-Mn-Ga films" *Physics Procedia*, 10 (2010) 162-167.

- M. Kohl, B. Krevet, **S.R. Yeduru**, Y. Ezer, A. Sozinov "A novel foil actuator using magnetic shape memory effect" *Smart Mater. Struct.,* 20 (2011) 094009.

- **S.R. Yeduru**, A. Backen, C. Kübel, D. Wang, S. Torsten, S. Fähler, L. Schultz and M. Kohl "Microstructure of freestanding epitaxial Ni-Mn-Ga films before and after variant reorientation" *Scripta Materialia*, (2012) DOI:10.1016/j.scriptamat.2012.01.004

- **S.R. Yeduru**, A. Backen, S. Fähler, L. Schultz and M. Kohl "Transformation behavior of freestanding epitaxial Ni-Mn-Ga films" *Journal of Alloys and Compounds,* (2012) in press DOI: 10.1016/j.jallcom.2012.02.151

- A. Backen, **S.R. Yeduru**, M. Kohl, L. Schultz and S. Fähler, "Ni-Mn-Ga films for magnetic shape memory alloy microactuators" *Advanced Engineering Materials* 14 (2012) 696-709 DOI: 10.1002/adem.201200069

### *Conference proceedings*

- F. Khelfaoui, **S.R. Yeduru**, M. Kohl, A. Mecklenburg, and R. Schneider "Magnetic field induced reorientation in single crystalline Ni-Mn-Ga foil actuators" *Proc. European Symposium on Martensitic Transformation* ESOMAT-2009, 07013 (2009), DOI:10.1051/esomat/200907013.
- M. Kohl, B. Krevet, **S.R. Yeduru**, Y. Ezer, A. Sozinov "A ferromagnetic shape memory foil actuator" *ASME Conf. Proc. 2010*, SMASIS 2010-3652 pp 41-48, DOI: 10.115/SMASIS2010-3652
- **S.R. Yeduru**, Y. Ezer, A. Sozinov, M. Kohl "Linear actuation based on a magnetic shape memory foil" *Actuator 10,* conference proceedings, (2010) pp 303-306.
- **S.R. Yeduru**, A. Backen, S. Fähler, A. Dienstel, L. Schultz, M. Kohl "Sacrificial layer technology- a route towards freestanding Ni-Mn-Ga films" *Actuator 10*, Conference proceedings, (2010) pp 765-768.

Schriften des Instituts für Mikrostrukturtechnik
am Karlsruher Institut für Technologie (KIT)

ISSN 1869-5183

Herausgeber: Institut für Mikrostrukturtechnik

Die Bände sind unter www.ksp.kit.edu als PDF frei verfügbar
oder als Druckausgabe zu bestellen.

Band 1 Georg Obermaier
Research-to-Business Beziehungen: Technologietransfer durch Kommunikation von Werten (Barrieren, Erfolgsfaktoren und Strategien). 2009
ISBN 978-3-86644-448-5

Band 2 Thomas Grund
Entwicklung von Kunststoff-Mikroventilen im Batch-Verfahren. 2010
ISBN 978-3-86644-496-6

Band 3 Sven Schüle
Modular adaptive mikrooptische Systeme in Kombination mit Mikroaktoren. 2010
ISBN 978-3-86644-529-1

Band 4 Markus Simon
Röntgenlinsen mit großer Apertur. 2010
ISBN 978-3-86644-530-7

Band 5 K. Phillip Schierjott
Miniaturisierte Kapillarelektrophorese zur kontinuierlichen Überwachung von Kationen und Anionen in Prozessströmen. 2010
ISBN 978-3-86644-523-9

Band 6 Stephanie Kißling
Chemische und elektrochemische Methoden zur Oberflächenbearbeitung von galvanogeformten Nickel-Mikrostrukturen. 2010
ISBN 978-3-86644-548-2

Schriften des Instituts für Mikrostrukturtechnik
am Karlsruher Institut für Technologie (KIT)

**ISSN 1869-5183**

| | |
|---|---|
| Band 7 | Friederike J. Gruhl<br>Oberflächenmodifikation von Surface Acoustic Wave (SAW) Biosensoren für biomedizinische Anwendungen. 2010<br>ISBN 978-3-86644-543-7 |
| Band 8 | Laura Zimmermann<br>Dreidimensional nanostrukturierte und superhydrophobe mikrofluidische Systeme zur Tröpfchengenerierung und -handhabung. 2011<br>ISBN 978-3-86644-634-2 |
| Band 9 | Martina Reinhardt<br>Funktionalisierte, polymere Mikrostrukturen für die dreidimensionale Zellkultur. 2011<br>ISBN 978-3-86644-616-8 |
| Band 10 | Mauno Schelb<br>Integrierte Sensoren mit photonischen Kristallen auf Polymerbasis. 2012<br>ISBN 978-3-86644-813-1 |
| Band 11 | Daniel Auernhammer<br>Integrierte Lagesensorik für ein adaptives mikrooptisches Ablenksystem. 2012<br>ISBN 978-3-86644-829-2 |
| Band 12 | Nils Z. Danckwardt<br>Pumpfreier Magnetpartikeltransport in einem Mikroreaktionssystem: Konzeption, Simulation und Machbarkeitsnachweis. 2012<br>ISBN 978-3-86644-846-9 |
| Band 13 | Alexander Kolew<br>Heißprägen von Verbundfolien für mikrofluidische Anwendungen. 2012<br>ISBN 978-3-86644-888-9 |

Schriften des Instituts für Mikrostrukturtechnik
am Karlsruher Institut für Technologie (KIT)

ISSN 1869-5183

Band 14 Marko Brammer
Modulare Optoelektronische Mikrofluidische Backplane. 2012
ISBN 978-3-86644-920-6

Band 15 Christiane Neumann
Entwicklung einer Plattform zur individuellen Ansteuerung von Mikroventilen und Aktoren auf der Grundlage eines Phasenüberganges zum Einsatz in der Mikrofluidik. 2013
ISBN 978-3-86644-975-6

Band 16 Julian Hartbaum
Magnetisches Nanoaktorsystem. 2013
ISBN 978-3-86644-981-7

Band 17 Johannes Kenntner
Herstellung von Gitterstrukturen mit Aspektverhältnis 100 für die Phasenkontrastbildgebung in einem Talbot-Interferometer. 2013
ISBN 978-3-7315-0016-2

Band 18 Kristina Kreppenhofer
Modular Biomicrofluidics - Mikrofluidikchips im Baukastensystem für Anwendungen aus der Zellbiologie. 2013
ISBN 978-3-7315-0036-0

Band 19 Ansgar Waldbaur
Entwicklung eines maskenlosen Fotolithographiesystems zum Einsatz im Rapid Prototyping in der Mikrofluidik und zur gezielten Oberflächenfunktionalisierung. 2013
ISBN 978-3-7315-0119-0

Band 20 Christof Megnin
Formgedächtnis-Mikroventile für eine fluidische Plattform. 2013
ISBN 978-3-7315-0121-3

Schriften des Instituts für Mikrostrukturtechnik
am Karlsruher Institut für Technologie (KIT)

**ISSN 1869-5183**

**Band 21** **Srinivasa Reddy Yeduru**
**Development of Microactuators Based on the Magnetic Shape Memory Effect.** 2013
ISBN 978-3-7315-0125-1